Models in Kinematics for Engineers

Sorin Vlase · Marin Marin · Andreas Öchsner ·
Maria Luminita Scutaru

Models in Kinematics for Engineers

Springer

Sorin Vlase
Institute of Solid Mechanics
of the Romanian Academy
Transilvania University of Braşov
Braşov, Romania

Marin Marin
Department of Mathematics and Computer Science
Transilvania University of Braşov
Braşov, Romania

Andreas Öchsner
Faculty of Mechanical Engineering
Esslingen University of Applied Sciences
Esslingen am Neckar, Baden-Württemberg
Germany

Maria Luminita Scutaru
Department of Mechanical Engineering
Transilvania University of Braşov
Braşov, Romania

ISBN 978-3-032-28112-8 ISBN 978-3-032-28113-5 (eBook)
https://doi.org/10.1007/978-3-032-28113-5

This Springer imprint is published by the registered company Springer Nature Switzerland AG
The registered company address is: Gewerbestrasse 11, 6330 Cham, Switzerland

Preface

The content of this textbook covers all the standard introductory topics in classical mechanics, for the main part the kinematics (the analysis of trajectories, velocities and accelerations of different points of a body). Starting from basic liaison conditions between different bodies existing in mechanics are analyzed the principal technical pairs of bodies that assure the technical contact in a multibody system. The main problems that may arise in engineering practice are analyzed, and numerous problems illustrate the presentation. As a theoretical field, classical mechanics is noted for its intrinsic beauty, richness, and depth as well as its coherence and elegance. This book aims to emphasize the harmony and beauty of classical mechanics. Because of the writers' extensive experience, their presentation style has been put to the test throughout many years of working with pupils. Advanced undergraduates and first-year graduate students interested in the engineering application of contemporary techniques in classical mechanics are the target audience for this textbook. To encourage a thorough comprehension of the topic, the writers make an effort to write in an understandable and methodical manner. The writers used practical examples to motivate and illustrate each subject in this section of mechanics and statics.

It is well known that classical mechanics, viewed as a theoretical discipline, possesses an inherent beauty, depth, and richness and presents coherence and elegance. This book tries to highlight this beauty and harmony that classical mechanics offers. The long experience of the authors means that the way of presentation is intensively tested in the decades of contact with students. The textbook is mainly addressed to advanced undergraduate and beginning graduate students who are interested in the engineering application of modern methods in classical mechanics. The authors try to use a clear and systematic style to promote a good understanding of the subject. For this part of mechanics, kinematics, the authors motivated and illustrated each concept, with worked examples. The goal of the book is to give a comprehensive overview of the basic ideas and methods of classical mechanics. The writers' many years of experience giving lectures and seminars served as the basis for the book. The majority of the issues are unique and will be helpful to both mechanics students and teachers.

Braşov, Romania Sorin Vlase
Braşov, Romania Marin Marin
Esslingen am Neckar, Germany Andreas Öchsner
Braşov, Romania Maria Luminita Scutaru

Competing Interests The authors have no competing interests to declare that are relevant to the content of this manuscript.

Contents

Chapter 1
Kinematics of the Material Point

1.1 Overview and Initial Definition

The following definitions will clarify the object of kinematics.

Kinematics: The branch of mechanics that deals with the study of the movement of bodies, independent of their masses and the causes that produce the movement [1].

Kinematik, Bewegungslehre, Phoronomie: Der Teil der Mechanik, in dem allein die Bewegung der Körper ohne Rücksicht auf die sie verursachenden Kräfte untersucht wird. Dabei hat die Kinematik rein geometrischen Charakter. Man unterscheidet die Kinematik des Punktes und des starren Körpers. Zur Darstellung aller Bewegungen dienen die Begriffe Bahn, Geschwinigkeit und Beschleunigung. Die Keplerschen Gesetze sind z. B. rein kinematische Gesetze. In der Technik ist die Kinematik ein wichtiger Teil der Getriebelehre [2].

Kinematics: Branch of physics and subdivision of classical mechanics, concerns the description of the motion of objects without considering the forces that cause or result from motion. It is an abstract study of motion that aims to provide a description of the spatial position of points in moving bodies, the rate at which the points are moving (velocity), and the rate at which their velocity is changing (acceleration). When the causative forces are disregarded, motion descriptions are possible only for points having constrained motion; i.e., moving on determinate paths. In unconstrained or free motion the forces determine the shape of the path [3].

Kinematics deals with the study of the motion of a point, systems of points and bodies from a geometric point of view (without considering their masses and the interactions between them). Of the three fundamental notions used in mechanics, space, time and mass, kinematics uses only the first two. For this reason it is also called the geometry of motion (phoronomy). The study of kinematics does not involve the fundamental theorems of mechanics but is indispensable for solving problems. It provides us with approach methodologies and a calculation technique adapted to the situations imposed by the theoretical development. The modern approach to

S. Vlase et al., *Models in Kinematics for Engineers*,
https://doi.org/10.1007/978-3-032-28113-5_1

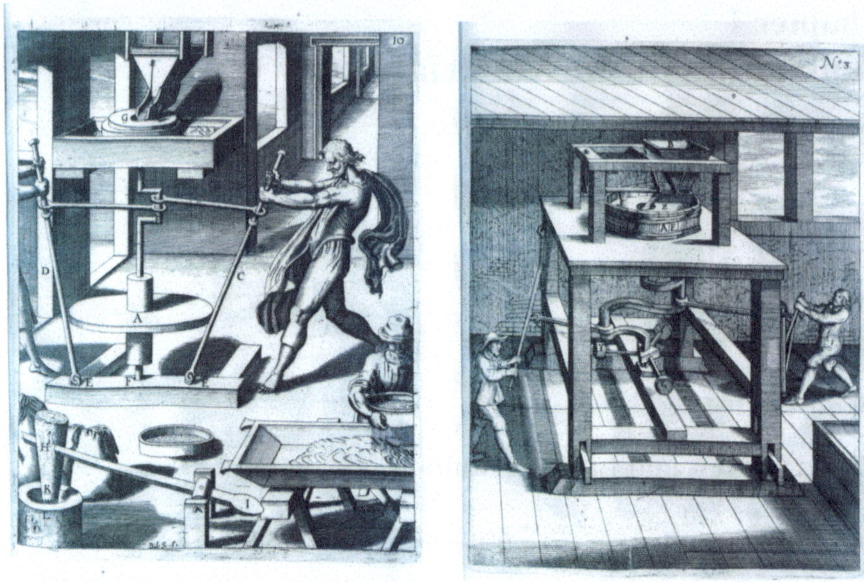

Fig. 1.1 Mechanism with bars [4, 5]. Transformation of the translation motion into rotation

mechanics, by using differential and geometric representations, emphasizes the main characteristics of kinematics.

Elements of kinematics were known since antiquity, but a special development of kinematics takes place starting from the renaissance period and continuing with the development of machines and mechanisms. The peak of kinematics can be considered to be in the nineteenth century, when the transmission motion was systematically studied.

Several drawings present, from a historical perspective, different types of mechanisms, known and used in antiquity and renaissance (Figs. 1.1, 1.2, 1.3, and 1.4).

1.2 The General Elements of Point Movement

1.2.1 Introduction

The study of kinematics follows the evolution of a point or systems of points in space and time. A rigid boddy is a system of material points, so it falls into the same category. We will begin the presentation of kinematics graded, studying first of all the simplest problem, that of the material point [6–8].

Let us assume a material point that has some movement in space. The position of a point in relation to a reference system can be expressed in different forms, but

a) b)

Fig. 1.2 **a** Variator with belts; **b** Worm transmission [4, 5]

a) b)

Fig. 1.3 **a** Slider crank mechanism; **b** Rack [4, 5]

a) b)

Fig. 1.4 **a** Motion transmission; **b** Gear and mechanism with ratchet [4, 5]

the most used is the vector representation. The position vector of the point relative to the origin O of a reference system will be a function $\vec{r} = \vec{r}(t)$. If this function is known, the movement of the point is completely determined. Physical considerations for the function to represent a real movement require that it is uniformly continuous (the point cannot jump from one position to another, it must travel through all the points of a continuous curve that joins the two positions), derivable (the second law of Newton involves the second derivative of the position vector) and finite in module (observable space is at finite distance). The fundamental quantities in the kinematics of the material point are those of trajectory, velocity and acceleration. When working with rigid bodies, the definition of angular displacements gives rise to the notions of angular velocity and angular acceleration.

1.2.2 Trajectory

Definition The geometric locus of the successive positions of the moving material point is called a trajectory.

The trajectory is generally a curve (infinite or closed) or an arc of a curve (example: pendulum motion). The finite character of the possibility of our observation means that in practice we focus our attention on certain interesting portions of the trajectory. By referring to a referential system, the position vector $\vec{r}$ can generally be defined by three scalar components. If, for example, we consider Cartesian coordinates (Fig. 1.5), then we can write:

$$\vec{r} = x\vec{i} + y\vec{j} + z\vec{k}, \tag{1.1}$$

where the coordinates are functions of time:

$$x = x(t);\ y = y(t);\ z = z(t). \tag{1.2}$$

If the time between $x(t)$, $y(t)$ and $z(t)$ is eliminated, the equation of the trajectory can be obtained in the form of the intersection of two surfaces.

In the case of the cylindrical coordinate system, the three functions that define the position vector are the polar radius ρ, the polar angle θ and the elevation z (Fig. 1.6).

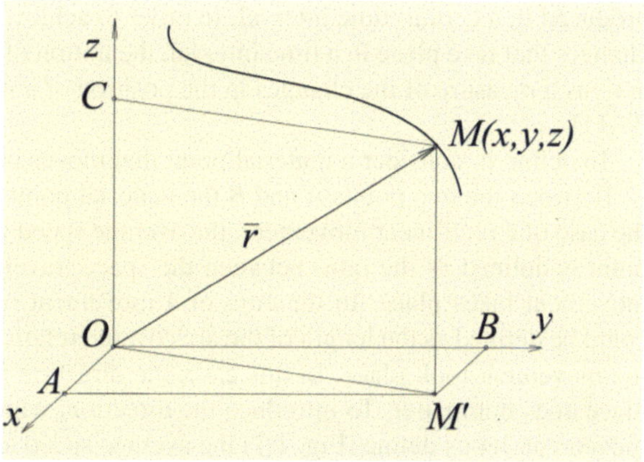

Fig. 1.5 Trajectory represented in a Cartesian coordinate space

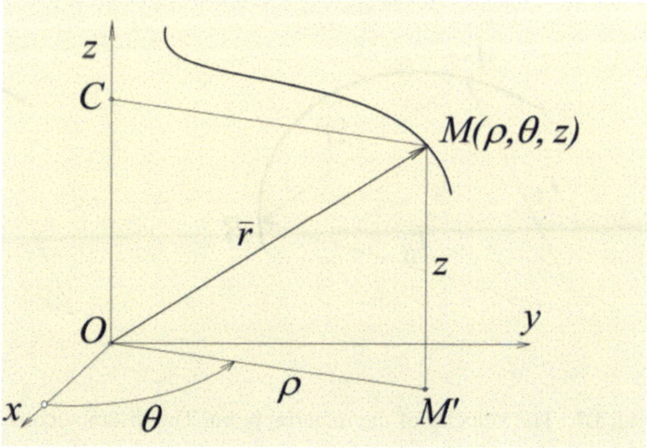

Fig. 1.6 Cylindrical coordinates

The following relationships:

$$\rho = \rho(t);\ \theta = \theta(t);\ z = z(t), \tag{1.3}$$

represent the parametric equations of the trajectory. The expression of kinematic quantities in different coordinate systems will be examined later.

1.2.3 Velocity

The notion of velocity quantitatively expresses a qualitative property of movement. Thus, in common language we have the notions of fast or slow, relative notions, by which we want to express how significant a change in a mechanical configuration is, produced in a certain time interval. In order to achieve an objective reporting of the changes that take place in a time interval, the notion of velocity is introduced, which assigns a measure of the changes in the position of a material point during this time [9–11].

To define it, consider a material point that moves on a curve (Fig. 1.7).

Between the two points A and B the material point can arrive faster or slower. In the case of a rectilinear movement, the average speed of the movement of a material point is defined as the ratio between the space traveled and the time in which the movement takes place. In the case of a movement on a curve, the average speed could be defined as the length of the arc covered relative to the time interval in which the movement took place. In this case, the direction in which the movement takes place does not matter. To introduce the direction, which is an important element of movement, let us define (Fig. 1.7) the average speed as a vector quantity:

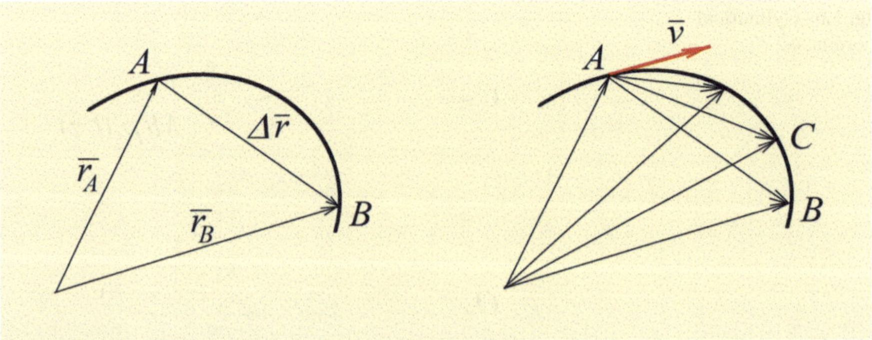

Fig. 1.7 The velocity of the material point. The instantaneous velocity is obtained decreasing the vector $\Delta\vec{r}$

$$\vec{v}_{\text{med}} = \frac{\Delta\vec{r}}{\Delta t} = \frac{\vec{r}_B - \vec{r}_A}{\Delta t} = \frac{\overrightarrow{AB}}{\Delta t}. \tag{1.4}$$

The average speed is a vector with the direction of $\overrightarrow{AB}$.

The average speed can incompletely characterize the (global) movement between two points because, for example, the first half of the road can be covered faster, and the second half slower. If we shorten the interval Δt so that the point goes from A to C, the average speed changes (see Fig. 1.7 right). This observation leads us to the idea of getting as close as possible to point A, making the interval Δt as small as possible. We can define in this case a quantity called the velocity of the material point at a given moment or more briefly, the velocity at point A:

$$\vec{v} = \lim_{\Delta t \to 0} \frac{\overrightarrow{AB}}{\Delta t} = \frac{d\vec{r}}{dt} = \dot{\vec{r}}. \tag{1.5}$$

In what follows, for the derivatives of a quantity with respect to time we will use Euler's notation, namely, the first derivative of the quantity a with respect to time will be denoted by $\dot{a}$, the second derivative by $\ddot{a}$, etc.

The velocity is tangent to the trajectory:

$$\vec{v} = \lim_{\Delta t \to 0} \frac{\overrightarrow{AB}}{\Delta t} = \lim_{\Delta t \to 0} \frac{\overrightarrow{AB}}{AB} \cdot \frac{AB}{archAB} \cdot \frac{archAB}{\Delta t} = \vec{\tau} \cdot 1 \cdot v = v \cdot \vec{\tau}, \tag{1.6}$$

where $\vec{\tau}$ is the unit vector tangent to the curve.

It was taken into account that:

$$\lim_{\Delta t \to 0} \frac{\overrightarrow{AB}}{AB} = \vec{\tau}; \qquad \lim_{\Delta t \to 0} \frac{AB}{archAB} = 1; \qquad \lim_{\Delta t \to 0} \frac{archAB}{\Delta t} = v. \tag{1.7}$$

The dimensional equation of the velocity is:

$$[v] = \frac{L}{T} = L \cdot T^{-1}.$$

The unit of measure for speed is the ratio of the unit of measure for length to the unit of measure for time. In SI units, speed is measured in meters per second (m/s). Tolerated units are kilometer per hour (km/h) and kilometer per second (km/s).

1.2.4 Projection of Movement on a Plane

Consider a material point M moving on a curve (C) and let M' be the orthogonal projection of the point on a given plane (P). Point M' will describe a trajectory (C') which is the projection of the curve (C) on the plane (Fig. 1.8).

We have the property: the velocity $\vec{v}'$ of point M' on the trajectory (C') is equal to the projection of the velocity of point M on the plane. Indeed, let two points M and M_1 be close enough so that we can consider the arc MM_1 equal to the arc MM_1. At time t the mobile is at point M and at time t_1 it is at M_1. The vector $\overrightarrow{MA}$ represents the average velocity of the material point on the curve (C) between the moments of time t and t_1. Its projection on the plane is $\overrightarrow{M'A'}$. We will show that this vector is the average velocity of the point M' on the curve (C'). Indeed:

(a) the parallels MM', M_1M_1', AA' determine on MA and $M'A'$ proportional segments:

$$\frac{M'A'}{M'M_1'} = \frac{MA}{MM_1}; \tag{1.8}$$

(b) the average velocity is:

$$v_{\text{med}} = MA = \frac{MM_1}{t_1 - t}. \tag{1.9}$$

Replacing MA by this value in the previous equality, we get:

$$\frac{M'A'}{M'M_1'} = \frac{MM_1}{MM_1(t_1 - t)} = \frac{1}{t_1 - t}, \tag{1.10}$$

where from:

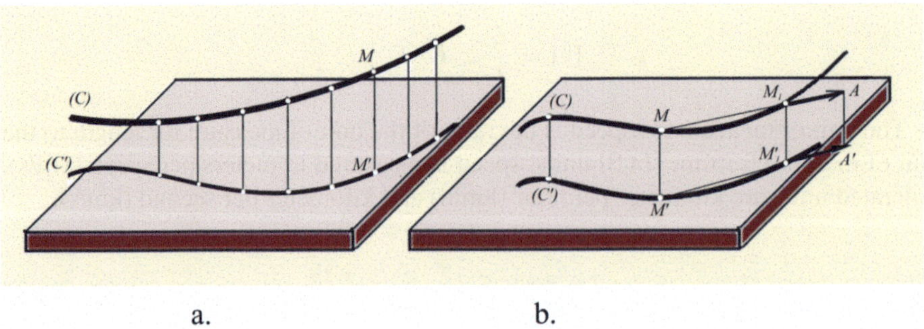

Fig. 1.8 Projection of movement on a plane. **a**. Projection of the trajectory; **b**. Projection of the velocity

$$M'A' = \frac{M'M_1'}{t_1 - t}, \tag{1.11}$$

which shows that $M'A'$ is the ratio between the space $M'M_1{}'$ and the time t_1-t required to travel this space. If t_1 tends to t, the vectors $\overrightarrow{MA}$ and $\overrightarrow{M'A'}$ tend to the vectors $\vec{v}$ and $\vec{v}'$, tangent respectively in M and M' to the curves (C) and (C'). The vector $\vec{v}'$ is therefore $\vec{v}$ projection on the given plane.

In an analogous way, the movement can be projected on an axis xx'. It can also be demonstrated in this case that the velocity of the projection of the point M on some axis xx' is equal to the projection of the velocity of the material point on the axis. This property applies to the study of harmonic oscillatory motion which can be treated as the projection of a material point moving on a circle, on a diameter of the circle.

1.2.5 Acceleration

When moving along a trajectory, the velocity can also vary. In a movement for example, at the beginning of the movement the material point can gain velocity and at the end of the movement it can lose velocity. Newton noticed that precisely this speed variation allows the construction of deterministic laws for the movement of bodies and links it to another important notion, namely force, through the law that bears his name. To characterize it, the notion of acceleration is introduced.

To define this quantity, let us move, through a translation, the velocity vector at point B to A (Fig. 1.9). The speed variation will be:

$$\Delta\vec{v} = \vec{v}_B - \vec{v}_A, \tag{1.12}$$

and its speed of variation is called average acceleration:

$$\vec{a}_{\text{med}} = \frac{\Delta\vec{v}}{\Delta t} = \frac{\vec{v}_B - \vec{v}_A}{\Delta t}. \tag{1.13}$$

As in the case of defining the velocity of a point, we can introduce the notion of instantaneous acceleration of the material point on the trajectory:

$$\vec{a} = \lim_{\Delta t \to 0} \frac{\Delta\vec{v}}{\Delta t} = \frac{d\vec{v}}{dt} = \frac{d^2\vec{r}}{dt^2} = \dot{\vec{v}} = \ddot{\vec{r}}. \tag{1.14}$$

The acceleration is a vector quantity whose expression is the derivative of velocity with respect to time. The dimensional equation of the acceleration will be written:

$$[a] = \frac{[v]}{[t]} = \frac{L}{T} \cdot \frac{1}{T} = L \cdot T^{-2}. \tag{1.15}$$

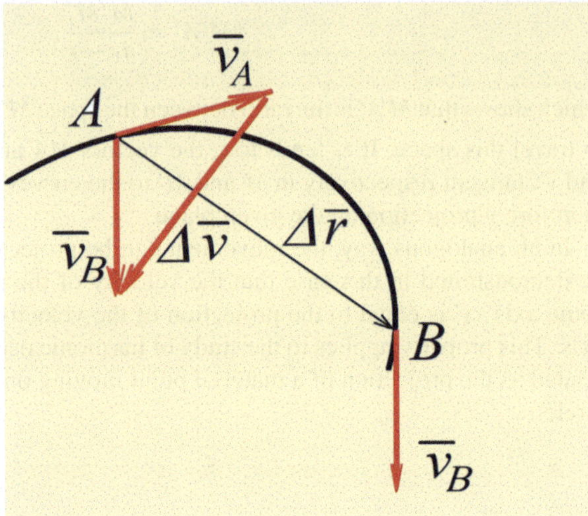

Fig. 1.9 Acceleration

The unit of measurement for acceleration is equal to the ratio of the unit of measurement for length to the square of the unit of measurement for time. In SI units, acceleration is measured in meters per second squared (m/s^2).

1.2.6 Higher Order Accelerations

Higher order accelerations can be obtained by differentiating more than twice $\vec{r}$ with respect to time. The third derivative with respect to the time of $\vec{r}$, so $\dddot{\vec{r}}$ it is called the acceleration of the second order, etc. The acceleration $\vec{a}$ is therefore the acceleration of the first order. In classical mechanics, the notion of acceleration, involved in Newton's second law $\vec{F} = m\vec{a}$, is sufficient. In the study of certain phenomena, higher-order accelerations are also taken into account, for phenomena with a very rapid variation in time. In our course we only limit to the first-order acceleration [12, 13].

1.2.7 Hodograph of the Velocity

A graphic representation, sometimes convenient in applications, can be obtained by introducing the notion of hodograph (in this case of velocity). Consider a material point that moves on a curve (C), (Fig. 1.10).

The velocities of different points are represented by applying them, through an equipole, to a point O, arbitrarily chosen in space (Fig. 1.10). If we move continuously

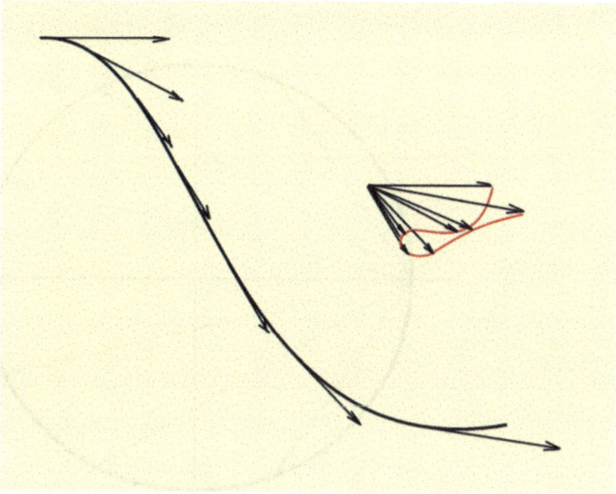

Fig. 1.10 Hodograph of the velocity

along the curve, the peak of the velocity vector will describe a curve (C') called the velocity hodograph. The arbitrarily chosen point O is called the pole of the hodograph. With this representation, the acceleration can be interpreted as the velocity of the extremity of the velocity vector on the hodograph (C').

For example, Fig. 1.11 shows the movement hodograph of circular motion with constant speed, in Fig. 1.12 the hodograph of the movement on the cycloid and in Fig. 1.13 the hodograph of the movement on the cylindrical helix and in the gravitational field.

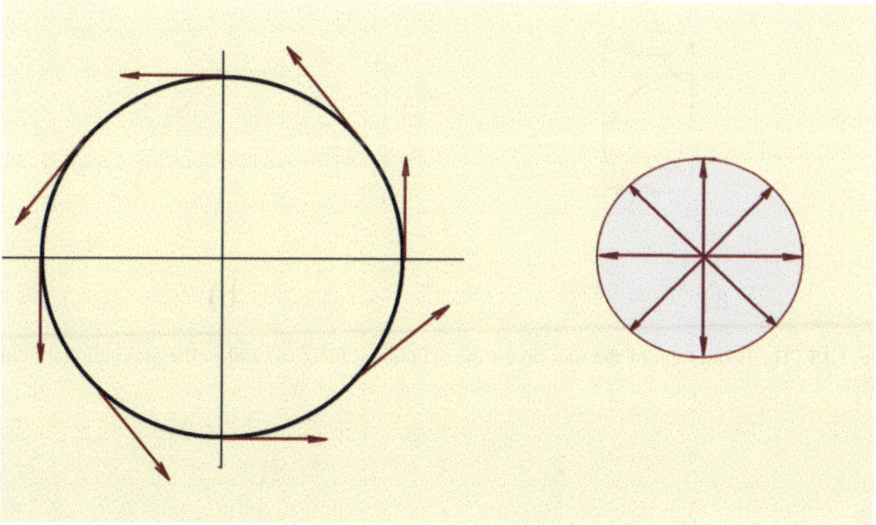

Fig. 1.11 Hodograph of circular motion

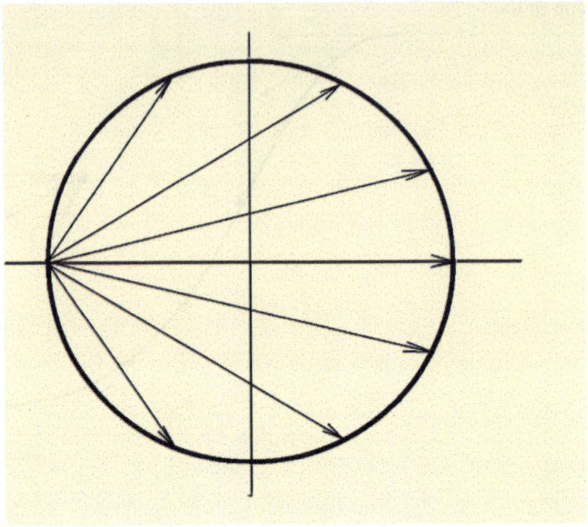

Fig. 1.12 Hodograph of the movement on the cycloid

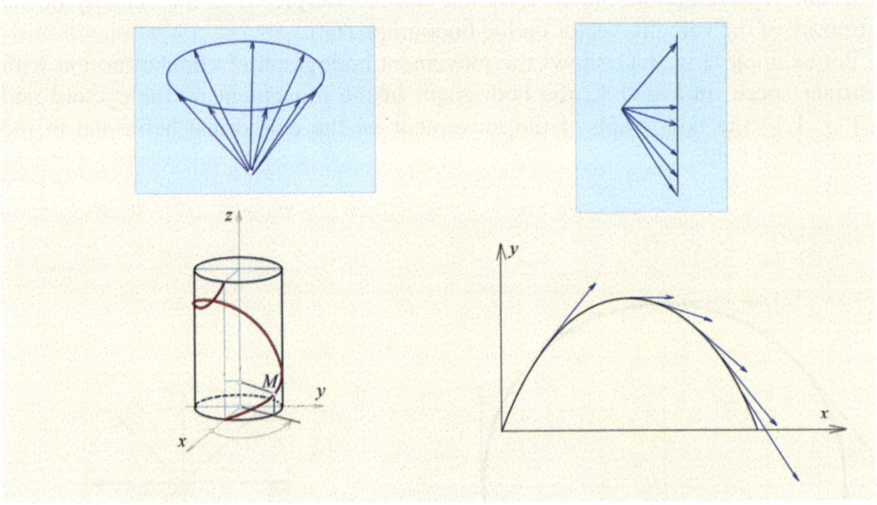

a) b)

Fig. 1.13 The hodograph of the motion on the cylindrical helix (**a**) and in the gravitational field (**b**)

1.3 Components of Velocity and Acceleration in Different Coordinate Systems

1.3.1 Cartesian Coordinate System

Expressions of Velocity and Acceleration

The position vector of the material point M (Fig. 1.14) $\vec{r} = \vec{r}(t)$, has the following components in the Cartesian reference system:

$$x = x(t);\ y = y(t);\ z = z(t), \tag{1.16}$$

which represent the parametric equations of the trajectory.

If time is eliminated, the trajectory equation can be obtained as the intersection of two surfaces:

$$\phi_1(x, y, z) = 0; \qquad \phi_2(x, y, z) = 0. \tag{1.17}$$

The velocity will be:

$$\vec{v} = \dot{x}\vec{i} + \dot{y}\vec{j} + \dot{z}\vec{k} = v_x\vec{i} + v_y\vec{j} + v_z\vec{k}, \tag{1.18}$$

with the value:

$$v = \sqrt{\dot{x}^2 + \dot{y}^2 + \dot{z}^2}. \tag{1.19}$$

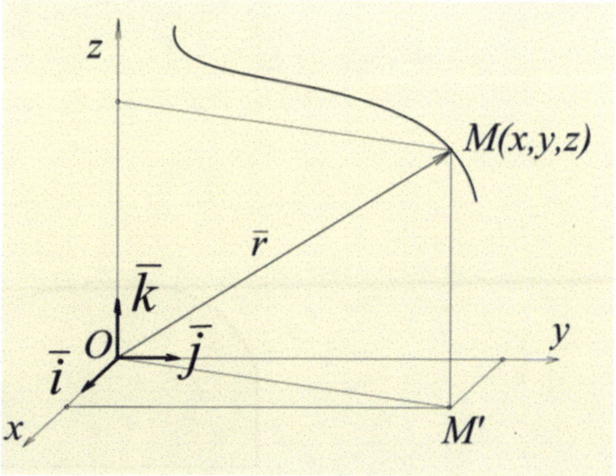

Fig. 1.14 Position vector of a material point in the Cartesian coordinate system

The acceleration is:

$$\vec{a} = \ddot{x}\vec{i} + \ddot{y}\vec{j} + \ddot{z}\vec{k} = a_x\vec{i} + a_y\vec{j} + a_z\vec{k}, \tag{1.20}$$

with the value:

$$a = \sqrt{\ddot{x}^2 + \ddot{y}^2 + \ddot{z}^2}. \tag{1.21}$$

In matrix form, we get:

$$\{r\} = \begin{Bmatrix} x \\ y \\ z \end{Bmatrix}; \quad \{v\} = \begin{Bmatrix} v_x \\ v_y \\ v_z \end{Bmatrix} = \begin{Bmatrix} \dot{x} \\ \dot{y} \\ \dot{z} \end{Bmatrix}; \quad \{a\} = \begin{Bmatrix} a_x \\ a_y \\ a_z \end{Bmatrix} = \begin{Bmatrix} \ddot{x} \\ \ddot{y} \\ \ddot{z} \end{Bmatrix};$$
$$v^2 = \{v\}^T\{v\} = \dot{x}^2 + \dot{y}^2 + \dot{z}^2; \quad a^2 = \{a\}^T\{a\} = \ddot{x}^2 + \ddot{y}^2 + \ddot{z}^2\,. \tag{1.22}$$

Applications. Motion on an Involute

Let us imagine a wire wound on a fixed curve (Fig. 1.15) tied at point *A*. Grasping the wire at point *B*, we unwind it while keeping it stretched. The geometric locus of the successive positions of point *B*, obtained in this operation, is called the involute. A curve has a family of involutes because any point on the wire can be chosen as point *B*. The involute is also called an unfolding curve (Latin evolvere means to unfold).

In what follows, the motion on the involute of the circle is studied. We propose to determine the analytical equation of the trajectory, the velocity and the acceleration of

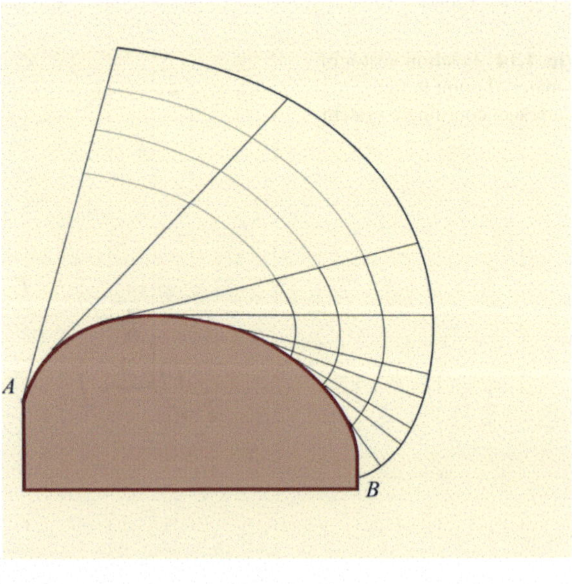

Fig. 1.15 Generation of an involute

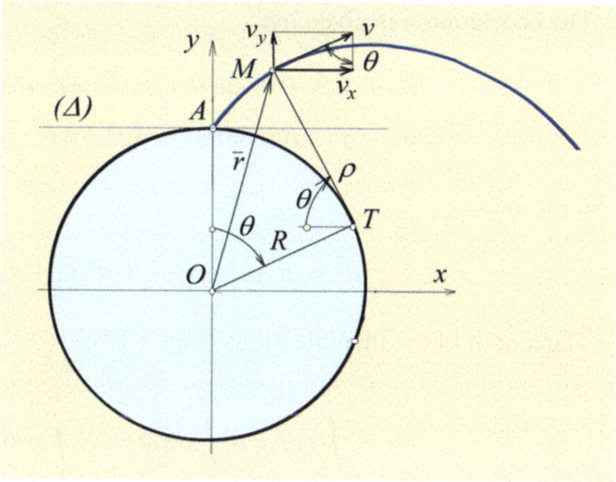

Fig. 1.16 The involute of the circle

a point of a straight line that rolls, without sliding, on a circle of radius R (Fig. 1.16). The angle made by the radius OT with the vertical is chosen as a kinematic parameter. We have:

$$TM = arcAT = R\theta. \tag{1.23}$$

$$\begin{aligned}\vec{r} &= \overrightarrow{OT} + \overrightarrow{TM} = (R\sin\theta\,\vec{i} + R\cos\theta\,\vec{j}) + (-R\theta\cos\theta\,\vec{i} + R\theta\sin\theta\,\vec{j}) \\ &= R(\sin\theta - \theta\cos\theta)\,\vec{i} + R(\cos\theta + \theta\sin\theta)\vec{j}.\end{aligned} \tag{1.24}$$

The parametric equations of the involute are therefore:

$$\begin{aligned}x &= R(\sin\theta - \theta\cos\theta); \\ y &= R(\cos\theta + \theta\sin\theta).\end{aligned} \tag{1.25}$$

The velocity components will be:

$$\begin{aligned}v_x &= \dot{x} = R\dot{\theta}(\cos\theta - \cos\theta + \theta\sin\theta) = R\theta\dot{\theta}\sin\theta; \\ v_y &= \dot{y} = R\dot{\theta}(-\sin\theta + \sin\theta + \theta\cos\theta) = R\theta\dot{\theta}\cos\theta,\end{aligned} \tag{1.26}$$

and the velocity is obtained from the relation:

$$v^2 = \dot{x}^2 + \dot{y}^2 = (R\theta\dot{\theta})^2, \tag{1.27}$$

where from:

$$v = \left|R\theta\dot{\theta}\right|. \tag{1.28}$$

The acceleration is given by:

$$a_x = \ddot{x} = R\theta\ddot{\theta}\sin\theta + R\dot{\theta}^2\sin\theta + R\theta\dot{\theta}^2\cos\theta$$
$$a_y = \ddot{y} = R\theta\ddot{\theta}\cos\theta + R\dot{\theta}^2\cos\theta - R\theta\dot{\theta}^2\sin\theta, \quad (1.29)$$

with the value:

$$a^2 = R^2(\theta^2\ddot{\theta} + \dot{\theta}^4 + \dot{\theta}^4\theta^2 + 4\theta\ddot{\theta}\dot{\theta}^2). \quad (1.30)$$

The length of the involute arc is (Fig. 1.17):

$$s = \int vdt = R\int_0^\alpha \theta\dot{\theta}dt = R\int_0^\alpha \theta d\theta = R\frac{\alpha^2}{2}, \quad (1.31)$$

(we suppose $\dot{\theta} \geq 0$). The angle made by the velocity vector with the vertical is given by:

$$tg\beta = \left|\frac{v_x}{v_y}\right| = tg\theta, \quad (1.32)$$

where from:

$$\beta = \theta. \quad (1.33)$$

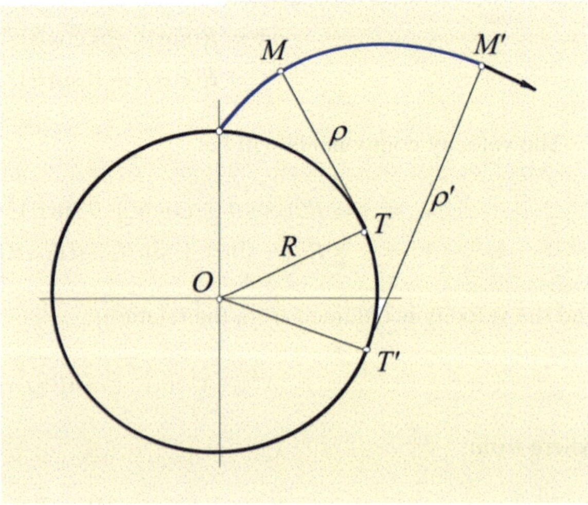

Fig. 1.17 The properties of the involute

It results:

- the velocity is perpendicular to TM (parallel to OT) (we have $\vec{v}\,\overrightarrow{TM} = 0$);
- has the value $\omega \cdot TM$, where we note $\dot{\theta} = \omega$:

$$v = \left|R\theta\dot{\theta}\right| = \dot{\theta} \cdot R\theta = \omega \cdot TM\,. \tag{1.34}$$

The point M on the line behaves as a function of the velocity (instantaneous) as if it had a circular motion on a circle of radius TM with the center in T, with the angular speed ω. If $\dot{\theta} = \omega = ct$, then:

$$\begin{aligned} a_x &= R\dot{\theta}^2(\sin\theta + \theta\cos\theta); \\ a_y &= R\dot{\theta}^2(\cos\theta - \theta\sin\theta); \end{aligned} \tag{1.35}$$

$$a^2 = a_x^2 + a_y^2 = (R\dot{\theta}^2)^2\left(1 + \theta^2\right) \tag{1.36}$$

where from:

$$a = (R\dot{\theta}^2)\sqrt{1+\theta^2} = \omega^2 \cdot OM\,. \tag{1.37}$$

Motion on the Cycloid

We propose to determine the trajectory, velocity and acceleration of a point on the periphery of a wheel of radius R that rolls without slipping on the horizontal plane (Fig. 1.18). We take as a variable parameter the angle θ. The position vector of the point M on the periphery of the wheel, after it has rolled by the angle θ, is given by the relation:

$$\begin{aligned} \vec{r} = \overrightarrow{OM} = \overrightarrow{OA} + \overrightarrow{AC} + \overrightarrow{CM} = R\theta\,\vec{i} \\ + R\vec{j} - R\sin\theta\,\vec{i} - R\cos\theta\,\vec{j} \end{aligned} \tag{1.38}$$

with the components:

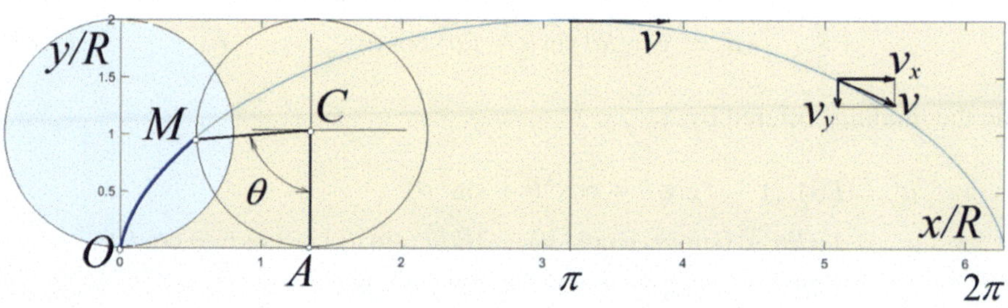

Fig. 1.18 The cycloid

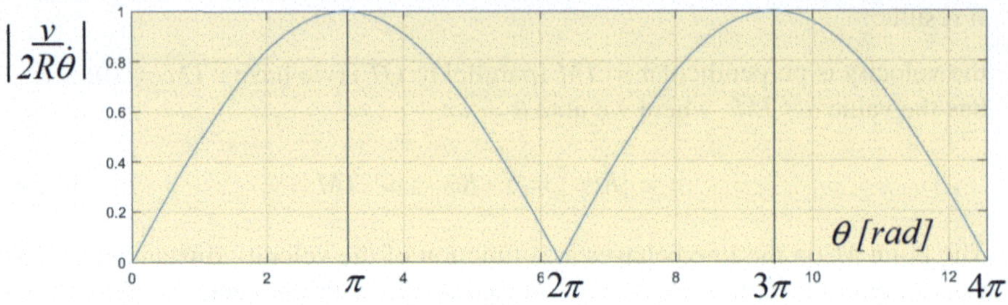

Fig. 1.19 Velocity modulus graph

$$\begin{aligned} x &= R(\theta - \sin\theta); \\ y &= R(1 - \cos\theta), \end{aligned} \tag{1.39}$$

which represent the parametric equations of the cycloid (the parameter being the angle θ). The velocity will be offered by the relations (Fig. 1.19):

$$\begin{aligned} v_x &= \dot{x} = R\dot{\theta}(1 - \cos\theta); \\ v_y &= \dot{y} = R\dot{\theta}\sin\theta, \end{aligned} \tag{1.40}$$

with the modulus calculated with:

$$\begin{aligned} v^2 &= \dot{x}^2 + \dot{y}^2 = (R\dot{\theta})^2(1 - 2\cos\theta + \cos^2\theta + \sin^2\theta) \\ &= 2(R\dot{\theta})^2(1 - \cos\theta) = 4(R\dot{\theta})^2\sin^2\frac{\theta}{2}; \end{aligned} \tag{1.41}$$

$$v = 2R\left|\dot{\theta}\sin\frac{\theta}{2}\right|. \tag{1.42}$$

The maximum value of the velocity is obtained for $\theta/2 = \pi/2$ so $\theta = \pi$.

The components of the acceleration of the point are given by the relations:

$$\begin{aligned} a_x &= \ddot{x} = R\ddot{\theta}(1 - \cos\theta) + R\dot{\theta}^2\sin\theta; \\ a_y &= \ddot{y} = R\ddot{\theta}\sin\theta + R\dot{\theta}^2\cos\theta, \end{aligned} \tag{1.43}$$

and the modulus offered by:

$$\begin{aligned} a^2 =& (R\ddot{\theta})^2(1 - 2\cos\theta + \cos^2\theta + \sin^2\theta) \\ &+ (R\dot{\theta}^2)^2(\sin^2\theta + \cos^2\theta) + 2R\ddot{\theta}\dot{\theta}^2\sin\theta(1 - \cos\theta + \cos\theta) \\ =& 2(R\ddot{\theta})^2(1 - \cos\theta) + (R\dot{\theta})^2 + 2R^2\ddot{\theta}\dot{\theta}^2\sin\theta \\ =& 4R^2\ddot{\theta}^2\sin^2\frac{\theta}{2} + 2R^2\ddot{\theta}\dot{\theta}^2\sin\theta + R^2\dot{\theta}^4, \end{aligned} \tag{1.44}$$

where from:

$$a = R\sqrt{4\ddot{\theta}^2 \sin\frac{\theta}{2} + 2\ddot{\theta}\dot{\theta}^2 \sin\theta + \dot{\theta}^4}. \tag{1.45}$$

The length of the cycloid arc is:

$$s = \int ds = \int \sqrt{\dot{x}^2 + \dot{y}^2}dt = \int vdt =$$

$$2R\int_0^\alpha \sin\frac{\theta}{2}\dot{\theta}dt = 2R\int_0^\alpha \sin\frac{\theta}{2}d\theta = 4R(1 - \cos\frac{\alpha}{2}); \tag{1.46}$$

$$s = 8R\sin^2\frac{\alpha}{2}. \tag{1.47}$$

During a complete roll of the circle, the space traveled by the material point is:

$$L = 8R\sin^2\frac{2\pi}{2} = 8R. \tag{1.48}$$

The angle made by the velocity vector with the horizontal is:

$$tg\beta = \frac{v_y}{v_x} = \frac{\dot{y}}{\dot{x}} = \frac{\sin\theta}{1 - \cos\theta} = \frac{2\sin\frac{\theta}{2}\cos\frac{\theta}{2}}{2\sin^2\frac{\theta}{2}}$$

$$= ctg\frac{\theta}{2} = tg\left(\frac{\pi}{2} - \frac{\theta}{2}\right), \tag{1.49}$$

so:

$$\beta = \frac{\pi}{2} - \frac{\theta}{2}. \tag{1.50}$$

(for $\theta \in [0; 2\pi]$) except when $\sin\theta/2 = 0$, $\theta/2 = k\pi$, $\theta = 2k\pi$. In this case: $\dot{x} = R\theta(1 - 1) = 0$; $\dot{y} = 0$, so v = 0. At the points $\theta = 2k\pi$, the angle made by the velocity with the horizontal shows a discontinuity:

$$tg\beta_1 = \lim_{\substack{\theta\to 2\pi \\ \theta<2\pi}} ctg\frac{\theta}{2} = -\infty; \quad \beta_1 = -\frac{\pi}{2}; \tag{1.51}$$

$$tg\beta_2 = \lim_{\substack{\theta\to 2\pi \\ \theta>2\pi}} ctg\frac{\theta}{2} = \infty; \quad \beta_2 = \frac{\pi}{2}; \tag{1.52}$$

The graph of the angle of inclination of the speed as a function of the angle θ of roll is given in Fig. 1.20.

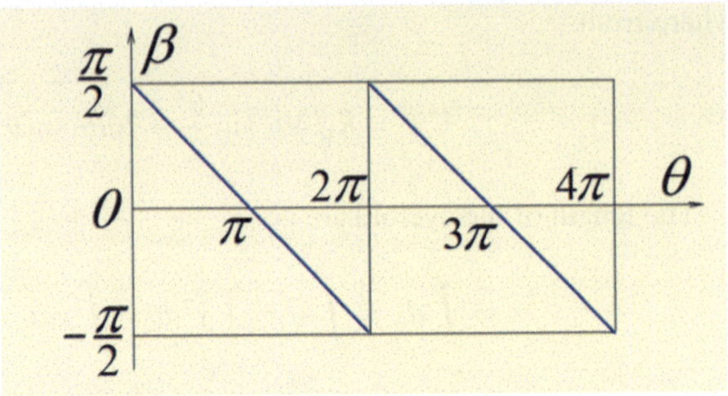

Fig. 1.20 The angle of inclination of the speed

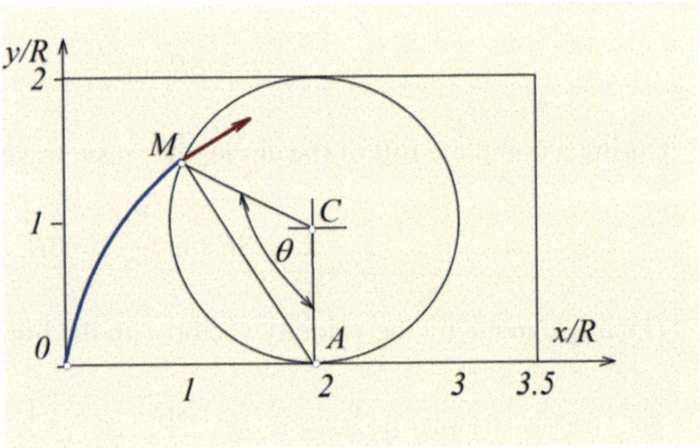

Fig. 1.21 The geometric interpretation of motion on a cycloid

If we consider that the wheel rolls at constant velocity $\dot{\theta} = ct = \omega$ then $\vec{v}$ and $\vec{a}$ have the following properties:

(a) the vector $\vec{v}$ is perpendicular to AM and equal in module with $\omega \cdot AM$. Indeed, $\beta = \pi/2 - (\pi - \theta)/2 = \theta/2$, $AM = 2R\sin(\theta/2)$ and then (Fig. 1.21):

$$\begin{aligned}\vec{v} \cdot \overrightarrow{AM} &= \left[R\theta(1-\cos\theta)\vec{i} + R\theta\sin\theta\vec{j}\right] \\ &\quad \left[-2R\sin\frac{\theta}{2}\cos\frac{\theta}{2}\vec{i} + 2R\sin^2\frac{\theta}{2}\vec{j}\right] \\ &= R\dot{\theta}^2\left[-(1-\cos\theta)\sin\frac{\theta}{2}\cos\frac{\theta}{2} + \sin\theta\sin^2\frac{\theta}{2}\right] = 0 \end{aligned} \tag{1.53}$$

$$v = 2R\omega\sin\frac{\theta}{2} = AM \cdot \omega. \tag{1.54}$$

It follows that the point M on the periphery of the wheel behaves as if the disc is rotating with angular speed ω around point A (the angular speed ω defines the time variation of the angle θ) (Fig. 1.21).

(b) The vector $\vec{a}$ has the direction of the MC radius and is equal with $\omega^2 \cdot MC$:

$$a_x = \ddot{x} = R\dot{\theta}^2 \sin\theta = R\omega^2 \sin\theta;$$
$$a_y = \ddot{y} = R\dot{\theta}^2 \cos\theta = R\omega^2 \cos\theta; \tag{1.55}$$

$$\overrightarrow{MC} = R\sin\theta\,\vec{i} + R\cos\theta\,\vec{j}, \tag{1.56}$$

so:

$$\vec{a} = a_x\vec{i} + a_y\vec{j} = (R\sin\theta\,\vec{i} + R\cos\theta\,\vec{j}) = \omega^2\overrightarrow{MC}. \tag{1.57}$$

The point M on the periphery of the wheel behaves, as if it is moving on the periphery of the circle of radius R, with constant speed (Fig. 1.21). If the material point is found inside a rolling circle, the shortened cycloid is obtained (Fig. 1.22) and if it is found on its exterior, forming a common body with the circle, the elongated cycloid is obtained (Fig. 1.23).

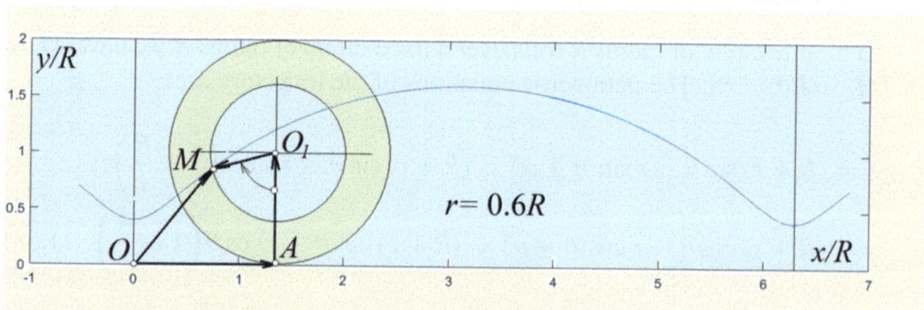

Fig. 1.22 Shortened cycloid

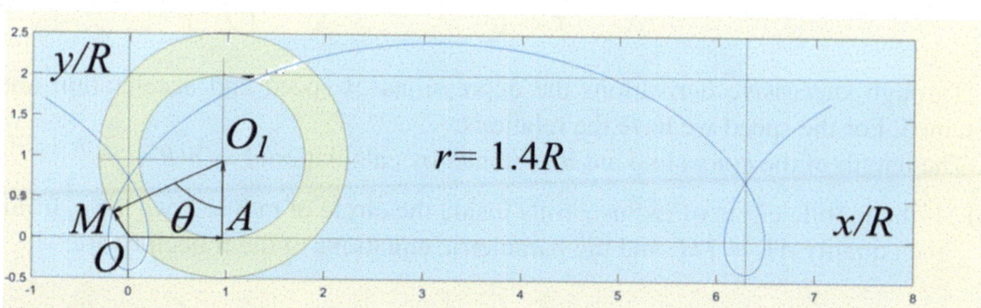

Fig. 1.23 Elongated cycloid

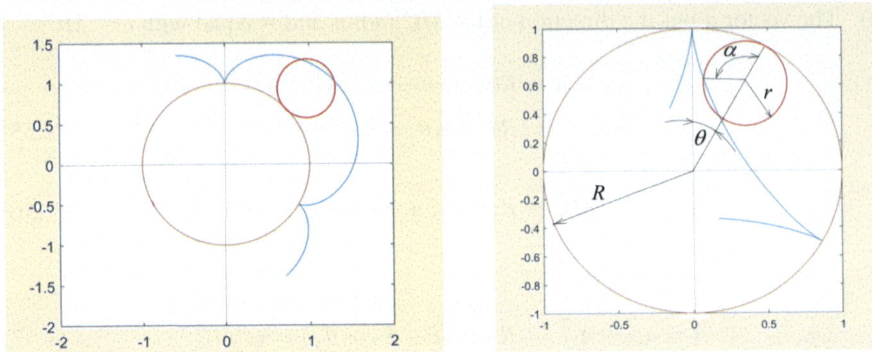

Fig. 1.24 Epicycloid and hypocycloid

Cycloidal curves. The curves studied above are called cycloidal curves. They are obtained by rolling, without friction, a mobile curve on a fixed curve (in the plane). If we are dealing with a mobile circle that rolls on the outside of a fixed circle, we get a curve called epicycloid, and if the mobile circle rolls on the inside of the fixed circle, we get a curve called hypocycloid (Fig. 1.24).

(a) If a mobile circle of radius R rolls over a fixed circle of radius R we have $AT = TM$, so $R\theta = r\alpha$. The parametric equations of the trajectory are:

$$x = (R+r)\sin\theta - r\sin(\theta+\alpha) = (R+r)\sin\theta - r\sin\theta\left(1+\frac{R}{r}\right);$$

$$y = (R+r)\cos\theta - r\cos(\theta+\alpha) = (R+r)\cos\theta - r\cos\theta\left(1+\frac{R}{r}\right). \quad (1.58)$$

$$v_x = \dot{x} = r\left(1+\frac{R}{r}\right)\dot{\theta}\left[\cos\theta - \cos\left(1+\frac{R}{r}\right)\theta\right];$$

$$v_y = \dot{y} = r\left(1+\frac{R}{r}\right)\dot{\theta}\left[-\sin\theta + \sin\left(1+\frac{R}{r}\right)\theta\right]. \quad (1.59)$$

Through successive derivations the expressions of speed and acceleration are obtained. For the speed we have the relations:

The length of the epicycloid arc is obtained by calculation, $l = 8(R+r)$.

(b) If the mobile circle of radius r rolls inside the circle of radius R we have, from the equality $AT = TM$, and the parametric equations of the trajectory are:

$$x = (R-r)\sin\theta - r\sin(-\theta+\alpha) = (R-r)\sin\theta - r\sin\theta\left(-1+\frac{R}{r}\right);$$

$$y = (R-r)\cos\theta + r\cos(-\theta+\alpha) = (R-r)\cos\theta + r\cos\theta\left(-1+\frac{R}{r}\right), \quad (1.60)$$

from where, through derivations, the velocity and acceleration are obtained:

$$v_x = \dot{x} = r\left(-1+\frac{R}{r}\right)\dot{\theta}\left[\cos\theta - \cos\left(-1+\frac{R}{r}\right)\theta\right];$$
$$v_y = \dot{y} = r\left(-1+\frac{R}{r}\right)\dot{\theta}\left[-\sin\theta - \sin\left(-1+\frac{R}{r}\right)\theta\right]. \tag{1.61}$$

The length of the hypocycloid arc, at a complete rotation, is $l = 8(R - r)$.

As an example of a cycloidal curve, consider a mobile circle that rolls on a fixed circle of the same radius R (Fig. 1.25). We will express the position vector of point M:

$$\begin{aligned}\vec{r} = \vec{OM} &= \vec{OA} + \vec{AB} + \vec{BM}\\ &= R\vec{i} + 2R(-\cos\theta\,\vec{i} + \sin\theta\,\vec{j}) + (R\cos 2\theta\,\vec{i} - R\sin 2\theta\,\vec{j})\\ &= (R - 2R\cos\theta + R\cos 2\theta)\vec{i} + (2R\sin\theta - R\sin 2\theta)\vec{j}\ .\end{aligned} \tag{1.62}$$

The observation that $ABMC$ is an isosceles trapezoid was used. It results:

$$x = R - 2R\cos\theta + R\cos 2\theta,\ y = 2R\sin\theta - R\sin 2\theta\ , \tag{1.63}$$

$$r^2 = x^2 + y^2 = 2R^2(3 - 4 - \cos\theta + \cos 2\theta) = 4R^2(1 - \cos\theta)^2;$$
$$r = 2R(1 - \cos\theta) = 2R(1 + \cos\alpha) = 4R\cos^2\frac{\alpha}{2}, \tag{1.64}$$

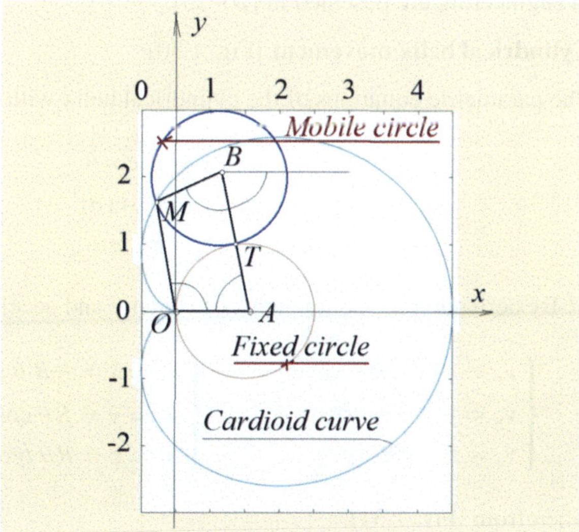

Fig. 1.25 Cardioid

that is, the equation of the cardioid. The velocity components will be:

$$\dot{x} = 2R\dot{\theta}(\sin\theta - \sin 2\theta)\ \dot{y} = 2R\dot{\theta}(\cos\theta - \cos 2\theta), \tag{1.65}$$

and the value:

$$v^2 = 4R^2\dot{\theta}^2(1 - \cos\theta) = 16R^2\dot{\theta}^2\sin^2\frac{\theta}{2}. \tag{1.66}$$

$$v = 4R\left|\dot{\theta}\sin\frac{\theta}{2}\right|; \tag{1.67}$$

$$\alpha + \theta = \pi;\ \dot{\alpha} = -\dot{\theta}. \tag{1.68}$$

The components of the acceleration are:

$$\begin{aligned} \ddot{x} &= 2R\ddot{\theta}(\sin\theta - \sin 2\theta) + 2R\dot{\theta}^2(\cos\theta - 2\cos 2\theta); \\ \ddot{y} &= 2R\ddot{\theta}(\cos\theta - \cos 2\theta) + 2R\dot{\theta}^2(-\sin\theta + 2\sin 2\theta). \end{aligned} \tag{1.69}$$

Figure 1.26a–d show epicycloids and hypocycloids for different values of the R/r ratio.

Spherical cycloidal curves are obtained by rotating, without sliding, one cone over another (in particular over a plane). Figures 1.27, 1.28 and 1.29 show some models from the Reuleux collection for viewing spherical cycloidal curves. They are obtained by rotating, without sliding, a circle (the base of the cone) on the circumference of another circle, the two circles being in different planes. Other technical application in engineering are pesented in [20–25].

Cylindrical helix movement (Fig. 1.30)

The parametric equations of the cylindrical helix with constant pitch are:

$$\begin{aligned} x &= R\cos\theta; \\ y &= R\sin\theta; \\ z &= R\theta\, tg\alpha. \end{aligned} \tag{1.70}$$

By derivation the components of velocity and acceleration are obtained:

$$\begin{cases} v_x = \dot{x} = -R\dot{\theta}\ \sin\theta; \\ v_y = \dot{y} = R\dot{\theta}\ \cos\theta; \\ v_z = \dot{z} = R\dot{\theta}\ tg\alpha\ . \end{cases} \quad \begin{cases} a_x = \ddot{x} = -R\ddot{\theta}\sin\theta - R\dot{\theta}^2\cos\theta; \\ a_y = \ddot{y} = R\ddot{\theta}\cos\theta - R\dot{\theta}^2\sin\theta; \\ a_z = \ddot{z} = R\ddot{\theta}\ tg\alpha\ . \end{cases} \tag{1.71}$$

wherefrom (Fig. 1.31):

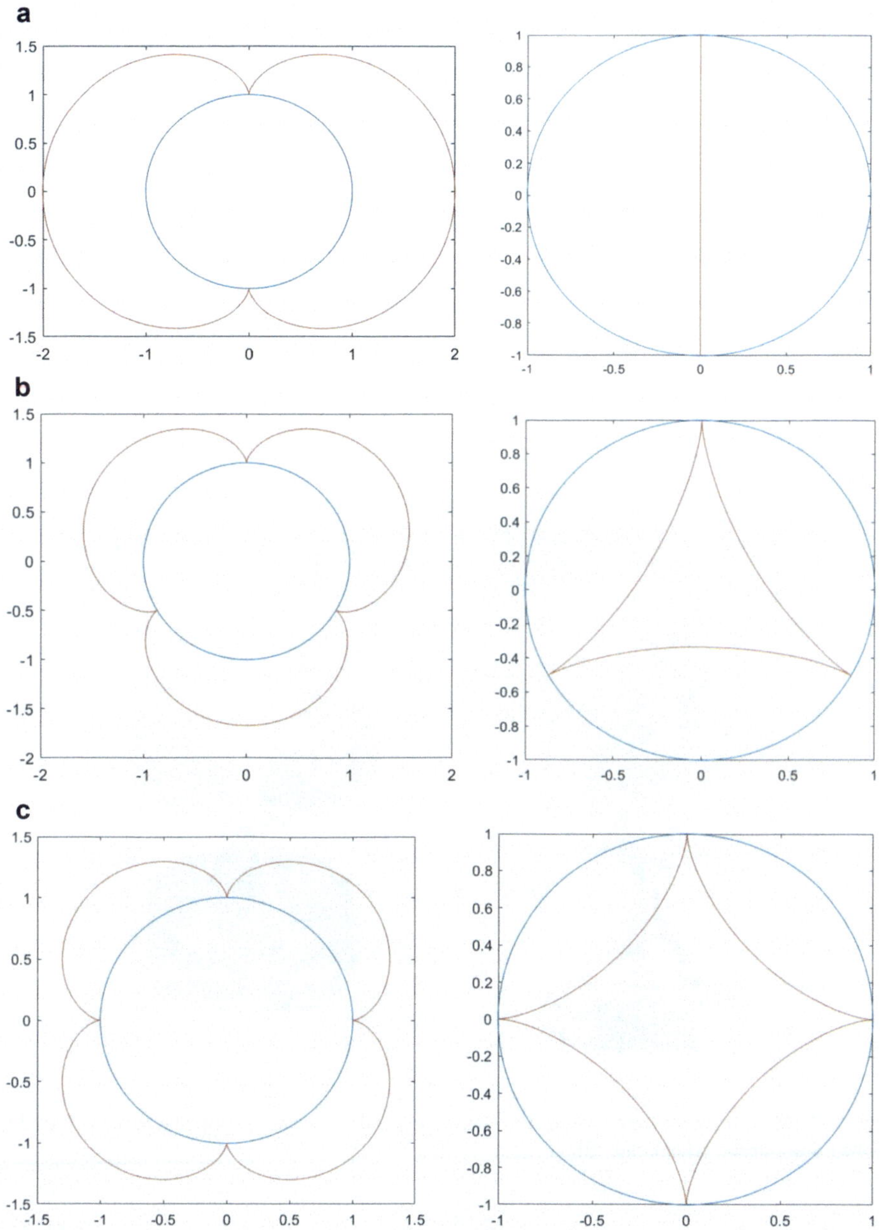

Fig. 1.26 **a** Epicycloid and hypocycloid for the case R/r = 2. **b** Epicycloid and hypocycloid for the case R/r = 3. **c** Epicycloid and hypocycloid for the case R/r = 4. **d** Epicycloid and hypocycloid for the case R/r = 5

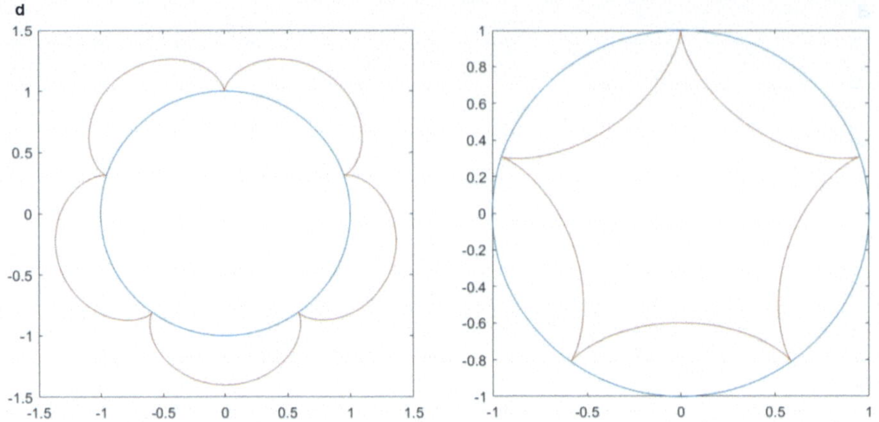

Fig. 1.26 (continued)

Fig. 1.27 Models for viewing spherical cycloidal curves. Left: plane rolling on a cone [14]; Right: Cone rolling inside other cone[15]

$$v^2 = \dot{x}^2 + \dot{y}^2 + \dot{z}^2 = \left(R\dot{\theta}\right)^2 \left(1 + tg^2\alpha\right) = R^2\,\dot{\theta}^2 \,/\, \cos^2\alpha; \tag{1.72}$$

$$v = R\,\left|\dot{\theta}\,\right| / \cos\alpha \quad ; \tag{1.73}$$

$$a^2 = R^2\ddot{\theta}^2 + R^2\,\dot{\theta}^4 + R^2\dot{\theta}^4 tg^2\alpha \ = R^2\left(\ddot{\theta}^2 / \cos^2\alpha + \dot{\theta}^4\right); \tag{1.74}$$

Fig. 1.28 Models for viewing spherical cycloidal curves. Left: Cone rolling outside a cone[16]; Right: Cone rolling inside other cone [17]

Fig. 1.29 Models for viewing spherical cycloidal curves. Left: plane rolling outside a cone [18]; Right: Cone rolling on a plane [19]

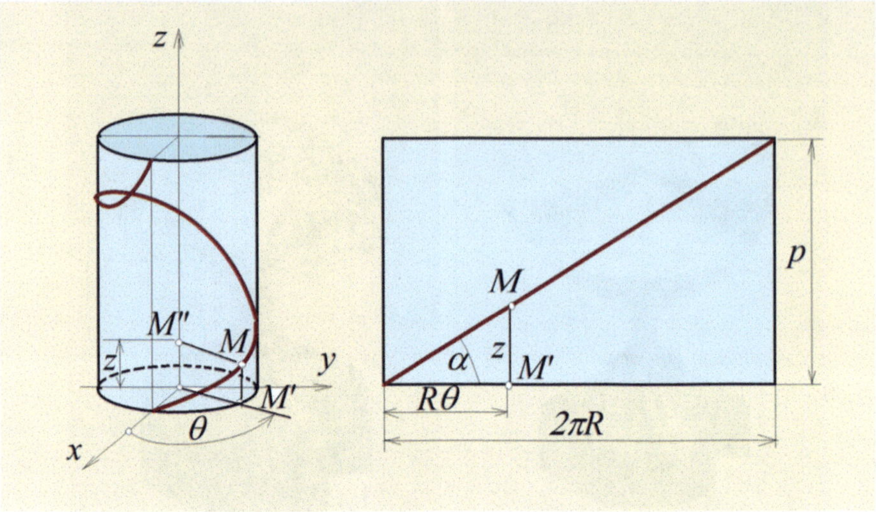

Fig. 1.30 Movement on the cylindrical helix

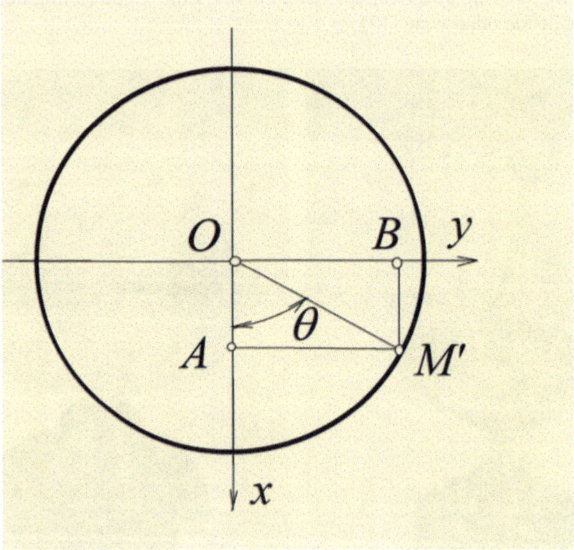

Fig. 1.31 Projection of the motion on a horizontal plane

$$a = R\sqrt{\left(\ddot{\theta}^2/\cos^2\alpha + \dot{\theta}^4\right)}\ . \tag{1.75}$$

The length of the helix arc is given by the relation:

$$s = \int ds = \int v\,dt = \frac{R}{\cos\alpha}\int_0^\beta \theta\,dt = \frac{R}{\cos\alpha}\,\beta. \tag{1.76}$$

1.3.2 Cylindrical Coordinate System

The position of a point M on the trajectory is defined by the coordinates θ (polar angle), ρ (polar radius) and z (elevation). Knowing the movement comes back to knowing the dependencies $\rho = \rho(t)$, $\theta = \theta(t)$ and $z = z(t)$ (Fig. 1.32).

A coordinate system related to the material point M is chosen in the following way: the unit vector $\vec{e}_\rho$ is parallel to the projection of the vector $\overrightarrow{OM}$ on the Oxy plane (it has the direction of the polar ray), the unit vector $\vec{e}_\theta$ is parallel to the Oxy plane and perpendicular to $\vec{e}_\rho$ so that the system formed by the unit vectors $\vec{e}_\rho$, $\vec{e}_\theta$ and $\vec{k}$ be right, $\vec{k}$ representing the direction of the Oz axis. In this case, the position vector of point M can be written:

$$\vec{r} = \rho\cdot\vec{e}_\rho + z\cdot\vec{k}\ . \tag{1.77}$$

The reference system determined by the three unit vectors is called the cylindrical reference system. It is a mobile system that moves with the point. As a result, let us see their variation of the unit vectors in relation to time. The unit vectors $\vec{e}_\rho$ and $\vec{e}_\theta$

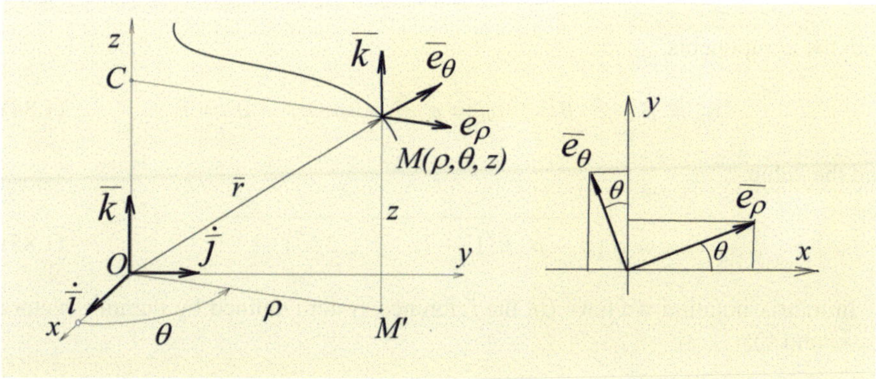

Fig. 1.32 Cylindrical coordinate system

are expressed as a function of θ through the relations:

$$\begin{aligned}\vec{e}_\rho &= \cos\theta\ \vec{i} + \sin\theta\ \vec{j};\\ \vec{e}_\theta &= -\sin\theta\ \vec{i} + \cos\theta\ \vec{j},\end{aligned} \tag{1.78}$$

from which we obtain, by derivation:

$$\begin{aligned}\dot{\vec{e}}_\rho &= \left(-\sin\theta\ \vec{i} + \cos\theta\ \vec{j}\right)\dot{\theta} = \dot{\theta}\cdot\vec{e}_\theta;\\ \dot{\vec{e}}_\theta &= -\left(\cos\theta\ \vec{i} + \sin\theta\ \vec{j}\right)\dot{\theta} = -\dot{\theta}\cdot\vec{e}_\rho;\\ \dot{\vec{k}} &= 0.\end{aligned} \tag{1.79}$$

The velocity, obtained by deriving the position vector, will be:

$$\vec{v} = \dot{\vec{r}} = \dot{\rho}\cdot\vec{e}_\rho + \rho\cdot\dot{\vec{e}}_\rho + \dot{z}\cdot\vec{k} = \dot{\rho}\cdot\vec{e}_\rho + \rho\cdot\dot{\theta}\cdot\vec{e}_\theta + \dot{z}\cdot\vec{k}, \tag{1.80}$$

with the components:

$$v_\rho = \dot{\rho}\ \ (radial); \qquad v_\theta = \rho\cdot\dot{\theta}\ \ (transversal); \ \ v_z = \dot{z}\ \ (axial), \tag{1.81}$$

and the value:

$$v = \sqrt{\dot{\rho}^2 + \rho^2\dot{\theta}^2 + \dot{z}^2}. \tag{1.82}$$

The acceleration, as the derivative of the velocity in relation to time, respectively as the second-order derivative of the position vector, will have the expression:

$$\begin{aligned}\vec{a} = \dot{\vec{v}} = \ddot{\vec{r}} &= \ddot{\rho}\cdot\vec{e}_\rho + \dot{\rho}\cdot\dot{\vec{e}}_\rho + \dot{\rho}\cdot\dot{\theta}\cdot\vec{e}_\theta + \rho\cdot\ddot{\theta}\cdot\vec{e}_\theta + \rho\cdot\dot{\theta}\cdot\dot{\vec{e}}_\theta + \ddot{z}\cdot\vec{k}\\ &= (\ddot{\rho} - \rho\cdot\dot{\theta}^2)\vec{e}_\rho + (\rho\cdot\ddot{\theta} + 2\dot{\rho}\cdot\dot{\theta})\vec{e}_\theta + \ddot{z}\cdot\vec{k},\end{aligned} \tag{1.83}$$

with the components:

$$a_\rho = \ddot{\rho} - \rho\cdot\dot{\theta}^2; \qquad a_\theta = \rho\cdot\ddot{\theta} + 2\dot{\rho}\cdot\dot{\theta}; \qquad a_z = \ddot{z}, \tag{1.84}$$

and the value:

$$a = \sqrt{\left(\ddot{\rho} - \rho\cdot\dot{\theta}^2\right)^2 + \left(\rho\cdot\ddot{\theta} + 2\dot{\rho}\dot{\theta}\right)^2 + \ddot{z}^2}. \tag{1.85}$$

In matrix notation we have (in the reference system defined by the unit vectors $\vec{e}_\rho$, $\vec{e}_\theta$ and $\vec{k}$):

$$\{r\} = \begin{Bmatrix} \rho \\ \theta \\ z \end{Bmatrix}; \quad \{v\} = \begin{Bmatrix} \dot{\rho} \\ \rho \cdot \dot{\theta} \\ \dot{z} \end{Bmatrix}; \quad \{a\} = \begin{Bmatrix} \ddot{\rho} - \rho \cdot \dot{\theta}^2 \\ \rho \cdot \ddot{\theta} + 2\dot{\rho} \cdot \dot{\theta} \\ \ddot{z} \end{Bmatrix}. \tag{1.86}$$

In the case of the movement of the material point on a cylinder $\rho = R$; $\dot{\rho} = 0$; $\ddot{\rho} = 0$, and the relations become:

$$\{r\} = \begin{Bmatrix} R \\ \theta \\ z \end{Bmatrix}; \quad \{v\} = \begin{Bmatrix} 0 \\ R \cdot \dot{\theta} \\ \dot{z} \end{Bmatrix}; \quad \{a\} = \begin{Bmatrix} -R \cdot \dot{\theta}^2 \\ R \cdot \ddot{\theta} \\ \ddot{z} \end{Bmatrix}. \tag{1.87}$$

The simplicity of the relations obtained justifies the use of these coordinates when the movement is made on a cylinder.

Application. Movement on the Cylindrical Propeller

The position vector can be written (see Fig. 1.33):

$$\vec{r} = R \cdot \vec{e}_\rho + R\theta \; tg\alpha \cdot \vec{k}. \tag{1.88}$$

Therefore, the velocity and acceleration result, through successive derivations:

$$\vec{v} = R\dot{\theta} \; \vec{e}_\theta + \dot{z}\,\vec{k} = R\dot{\theta} \left(\vec{e}_\theta + \vec{k} \; tg\alpha\right); \tag{1.89}$$

$$\vec{a} = -R\dot{\theta}^2 \,\vec{e}_\rho + R\ddot{\theta}\,\vec{e}_\theta + R\ddot{\theta}\, tg\alpha\,\vec{k}. \tag{1.90}$$

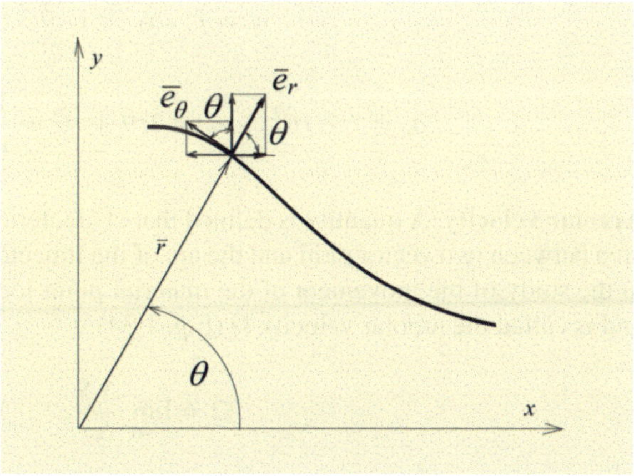

Fig. 1.33 Polar coordinate system

1.3.3 Polar Coordinate System

It is a particular case of the cylindrical coordinate system, when the movement takes place in the plane ($z = 0$) (Fig. 1.33).

The unit vectors of the plane coordinate system are $\vec{e}_\rho$ and $\vec{e}_\theta$ placed perpendicularly, so that they form a straight system. The mobile reference system moves with the material point. The derivatives of the unit vectors will be (they are obtained in the same way as in the case of cylindrical coordinates):

$$\dot{\vec{e}}_r = \dot{\theta} \cdot \vec{e}_\theta; \qquad \dot{\vec{e}}_\theta = -\dot{\theta} \cdot \vec{e}_r. \tag{1.91}$$

We can write:

$$\vec{r} = r \cdot \vec{e}_r. \tag{1.92}$$

In this case, the relationships that define the velocity and acceleration of the material point are:

$$\vec{v} = \dot{r} \cdot \vec{e}_r + r \cdot \dot{\theta} \cdot \vec{e}_\theta; \quad v = \sqrt{\dot{r}^2 + r^2\dot{\theta}^2}; \tag{1.93}$$

$$\vec{a} = \left(\ddot{r} - r\dot{\theta}^2\right)\vec{e}_r + \left(r\ddot{\theta} + 2\dot{r}\dot{\theta}\right)\vec{e}_\theta; \quad a = \sqrt{\left(\ddot{r} - r\dot{\theta}^2\right)^2 + \left(r\ddot{\theta} + 2\dot{r}\dot{\theta}\right)^2}. \tag{1.94}$$

The components of speed and acceleration, corresponding to the two unit vector of the coordinate system, can be written:

$$v_r = \dot{r}; \quad v_\theta = r \cdot \dot{\theta}; \tag{1.95}$$

$$a_r = \ddot{r} - r\dot{\theta}^2; \quad a_\theta = 2\dot{r}\dot{\theta} + r\ddot{\theta} = \frac{1}{r} \cdot \frac{d\left(r^2\dot{\theta}\right)}{dt}. \tag{1.96}$$

Areolar velocity. A quantity is defined that characterizes the increase in the sectoral area between two vector radii and the arc of the trajectory. This quantity is important in the study of the movement of the material point located in the central force field and is called the areolar velocity Ω (Fig. 1.34):

$$\Omega = \lim_{t \to 0} \frac{\Delta S}{\Delta t}. \tag{1.97}$$

From the figure it can be seen that Ω must check the relations:

$$\frac{r^2 \Delta \theta}{\Delta t} < \frac{\Delta S}{\Delta t} < \frac{(r + \Delta r)^2 \Delta \theta}{\Delta t}, \tag{1.98}$$

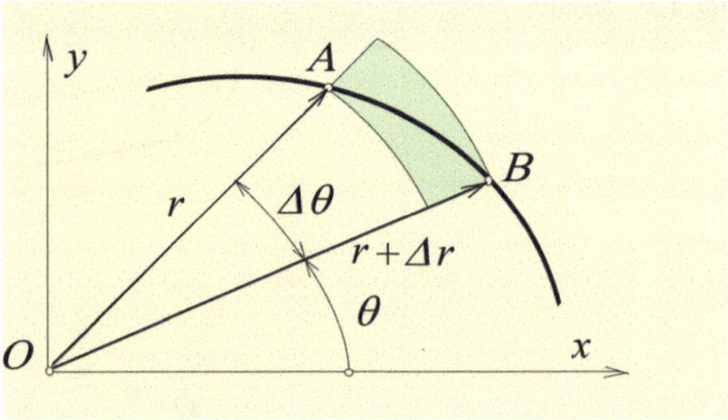

Fig. 1.34 Areolar velocity

in case $\Delta r > 0$. Passing to the limit results in:

$$\lim_{\Delta t \to 0} \frac{r^2 \Delta \theta}{\Delta t} \leq \lim_{\Delta t \to 0} \frac{\Delta S}{\Delta t} \leq \lim_{\Delta t \to 0} \frac{(r + \Delta r)^2 \Delta \theta}{\Delta t}, \tag{1.99}$$

where from: $\Omega = \frac{r^2 \dot{\theta}}{2}$. If $\Delta r < 0$ the reasoning is analogous. The notion of areolar velocity as well as its expression that were defined in the case of planar movement remain valid also in the case of the movement of the material point on a curve in space.

1.3.4 Natural Coordinate System (Frenet)

Expressions of Velocity and Acceleration

We assume that a material point moves on a curve (C) whose equation is known. A reference point A is taken on the curve from which the length of the arc of the curve is measured. The position of the material point can be defined in this case by the length of the arc of the curve s defined as a function of time $s = s(t)$ (equation of motion) (Fig. 1.35).

We attach to point M the mobile coordinate system with unit vectors $\vec{\tau}$ (following the direction of the tangent, in the direction of the increase of s), $\vec{\nu}$ in the direction normal to the curve (the main normal, oriented towards the concavity of the curve) and $\vec{\beta}$, the binormal versor given by the relation $\vec{\beta} = \vec{\tau} \times \vec{\nu}$. The considered trihedron is called Frenet's trihedron. There is the relationship:

$$\vec{v} = \dot{\vec{r}} = \frac{d\,\vec{r}}{d\,s} \cdot \frac{d\,s}{d\,t} = \dot{s} \cdot \vec{\tau} = v \cdot \vec{\tau}, \tag{1.100}$$

where $\frac{d\,\vec{r}}{d\,s} = \vec{\tau}$.

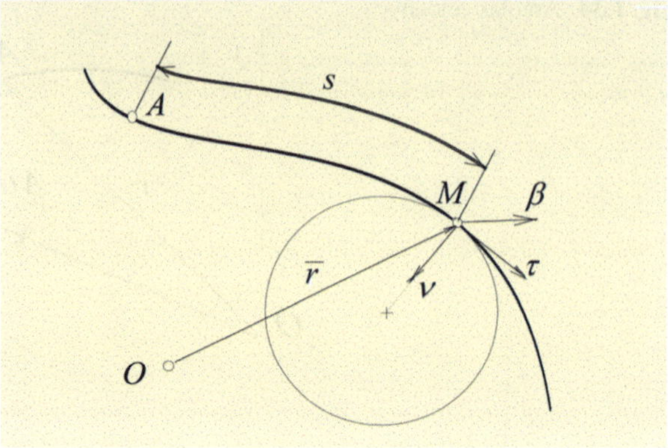

Fig. 1.35 Natural coordinate system

The components of the velocity on the axes of Frenet's trihadron are:

$$v_\tau = \dot{s}; \qquad v_\nu = 0; \qquad v_\beta = 0. \tag{1.101}$$

For the acceleration we have the relation:

$$\vec{a} = \frac{d\,\vec{v}}{d\,t} = \ddot{s} \cdot \vec{\tau} + \dot{s} \cdot \dot{\vec{\tau}}, \tag{1.102}$$

but $\dot{\vec{\tau}} = \frac{d\,\vec{r}}{d\,s} \cdot \frac{d\,s}{d\,t} = \dot{s} \cdot \frac{\vec{\nu}}{\rho}$. The relationship $\frac{d\,\vec{r}}{d\,s} = \frac{\vec{\nu}}{\rho}$ was used (Frenet's first formula). Then:

$$\vec{a} = \dot{v} \cdot \vec{\tau} + \frac{v^2}{\rho} \cdot \vec{\nu}. \tag{1.103}$$

The acceleration components will be:

$$a_\tau = \dot{v}; \quad a_\nu = \frac{v^2}{\rho}; \quad a_\beta = 0. \tag{1.104}$$

If $a_\tau = 0$ it results $v = ct$ and the movement is a uniform rectilinear movement. If a_τ and v have the same meaning, the movement is called accelerated ($a_\tau = ct.$, uniformly accelerated), and if a_τ and v have opposite signs, the movement is called delayed or slowed down ($a_\tau = ct.$, uniformly slowed down).

The normal acceleration is always directed along the normal since $a_\nu > 0$. If $a_\nu = 0$, $v = ct$ and $1/\rho = 0$, the motion is rectilinear and uniform.

The acceleration value is:

$$a = \sqrt{\dot{v}^2 + \frac{v^4}{\rho^2}}. \tag{1.105}$$

The Radius of Curvature of the Trajectory

From the formula written previously, the radius of curvature of the trajectory can be determined as follows:

$$\rho = \frac{v^2}{\sqrt{a^2 - \dot{v}^2}}. \qquad (1.106)$$

Another possibility to determine the radius of curvature follows the path used in differential geometry. Writing $\vec{v} = v\,\vec{\tau}; \quad \vec{a} = a_\tau\,\vec{\tau} + a_\nu\,\vec{\nu}$, it results:

$$\vec{v} \times \vec{a} = v\,a_\nu\,\vec{\tau} \times \vec{\nu} = \frac{v^3}{\rho}\,\vec{\beta}. \qquad (1.107)$$

So $|\vec{v} \times \vec{a}| = \frac{v^3}{\rho}$ and the radius of curvature will be:

$$\rho = \frac{v^3}{|\vec{v} \times \vec{a}|}. \qquad (1.108)$$

In Cartesian coordinates we have:

$$\vec{v}\,x\,\vec{a} = (\dot{y}\ddot{z} - \dot{z}\ddot{y})\vec{i} + (\dot{z}\ddot{x} - \dot{x}\ddot{z})\vec{j} + (\dot{x}\ddot{y} - \dot{y}\ddot{x})\vec{k}, \qquad (1.109)$$

where from:

$$\rho = \frac{(\dot{x}^2 + \dot{y}^2 + \dot{z}^2)^{3/2}}{\sqrt{(\dot{y}\ddot{z} - \dot{z}\ddot{y})^2 + (\dot{z}\ddot{x} - \dot{x}\ddot{z})^2 + (\dot{x}\ddot{y} - \dot{y}\ddot{x})^2}}. \qquad (1.110)$$

We have too:

$$\vec{v} \cdot \vec{a} = va^\tau, \qquad (1.111)$$

so:

$$a^\tau = \frac{\vec{v} \cdot \vec{a}}{v} = \frac{\dot{x}\ddot{x} + \dot{y}\ddot{y} + \dot{z}\ddot{z}}{\sqrt{\dot{x}^2 + \dot{y}^2 + \dot{z}^2}}, \qquad (1.112)$$

and:

$$a^\nu = \frac{v^2}{\rho} = \sqrt{\frac{(\dot{y}\ddot{z} - \dot{z}\ddot{y})^2 + (\dot{z}\ddot{x} - \dot{x}\ddot{z})^2 + (\dot{x}\ddot{y} - \dot{y}\ddot{x})^2}{\dot{x}^2 + \dot{y}^2 + \dot{z}^2}}. \qquad (1.113)$$

Examples Let us calculate the radius of curvature of the cycloid, the involute, the cylindrical helix and the cardioid.

The relation $\rho = \frac{v^2}{\sqrt{a^2-\dot{v}^2}}$ will be used and considered $\dot{\theta} = ct$ (since the radius of curvature must not depend on time). It is obtained:

(i) for the cycloid:

$$v = 2R\dot{\theta}\left|\sin\tfrac{\theta}{2}\right|,\ \dot{v} = R\dot{\theta}^2\left|\cos\tfrac{\theta}{2}\right|; \tag{1.114}$$

$$a = R\dot{\theta}^2; \tag{1.115}$$

$$\rho^2 = \frac{16R^4\dot{\theta}^4\sin^4\frac{\theta}{2}}{R^2\dot{\theta}^4 - 4R^2\dot{\theta}^4\cos^2\frac{\theta}{2}} = 16R^2\sin^2\frac{\theta}{2}. \tag{1.116}$$

(ii) for the involute:

$$v = R\left|\theta\dot{\theta}\right|; \tag{1.117}$$

$$\dot{v} = R\dot{\theta}^2; \tag{1.118}$$

$$a^2 = R^2\dot{\theta}^2(1+\theta^2); \tag{1.119}$$

$$\rho^2 = \frac{R^4\theta^4\dot{\theta}^4}{R^2(\dot{\theta}^4 + \dot{\theta}^4\theta^2 - \dot{\theta}^4)} = R^2\theta^2; \tag{1.120}$$

$$\rho = R\theta. \tag{1.121}$$

(iii) for the cylindrical helix:

$$v = \frac{R\dot{\theta}}{\cos\alpha}; \tag{1.122}$$

$$\dot{v} = 0; \tag{1.123}$$

$$a = R\dot{\theta}^2; \tag{1.124}$$

$$\rho = \frac{R}{\cos^2\alpha}. \tag{1.125}$$

(iv) for the cardioid:

$$\ddot{x} = 2R\dot{\theta}^2(\cos\theta - 2\cos 2\theta);\ \ddot{y} = 2R\dot{\theta}^2(-\sin\theta + 2\sin 2\theta); \tag{1.126}$$

$$a^2 = 4R^2\dot{\theta}^4(1 + 4 - 4\cos\theta) = 4R^2\dot{\theta}^2(5 - 4\cos\theta); \tag{1.127}$$

$$\dot{v} = 2R\dot{\theta}^2 \cos\frac{\theta}{2}; \tag{1.128}$$

$$\begin{aligned}\rho^2 &= \frac{256R^4\dot{\theta}^4 \sin^4\frac{\theta}{2}}{4R^2\dot{\theta}^4(5 - 4\cos\theta) - 4R^2\dot{\theta}^4\cos^2\frac{\theta}{2}}\\ &= \frac{64R^2\sin^4\frac{\theta}{2}}{4(1-\cos\theta) + \sin^2\frac{\theta}{2}} = \frac{64R^2\sin^4\frac{\theta}{2}}{8\sin^2\frac{\theta}{2} + \sin^2\frac{\theta}{2}} = \frac{64R^2\sin^2\frac{\theta}{2}}{9}\end{aligned} \tag{1.129}$$

From where:

$$\rho = \frac{8}{3}R\sin\frac{\theta}{2}. \tag{1.130}$$

1.3.5 Spherical Coordinate System

The position of the material point is determined by the size of the vector radius r, the angle θ (azimuth) and the angle ϕ (longitude). The movement of the material point is determined if the following dependencies are known:

$$r = r(t); \quad \theta = \theta(t); \quad \phi = \phi(t). \tag{1.131}$$

The unit vectors $\vec{e}_\rho$, $\vec{e}_\theta$ and $\vec{e}_\phi$ vertices represent the unit vectors of a right triorthogonal coordinate system (Fig. 1.36). The expressions of the three unit vectors are:

$$\begin{aligned}\vec{e}_\rho &= \sin\theta\cos\phi\,\vec{i} + \sin\theta\sin\phi\,\vec{j} + \cos\theta\,\vec{k};\\ \vec{e}_\theta &= \cos\theta\cos\phi\,\vec{i} + \cos\theta\sin\phi\,\vec{j} - \sin\theta\,\vec{k};\\ \vec{e}_\phi &= -\sin\phi\,\vec{i} + \cos\phi\,\vec{j},\end{aligned} \tag{1.132}$$

$$\{e_\rho\} = \begin{Bmatrix}\sin\theta\cos\phi\\ \sin\theta\sin\phi\\ \cos\theta\end{Bmatrix}; \quad \{e_\theta\} = \begin{Bmatrix}\cos\theta\cos\phi\\ \cos\theta\sin\phi\\ -\sin\theta\end{Bmatrix}; \quad \{e_\phi\} = \begin{Bmatrix}-\sin\phi\\ \cos\phi\\ 0\end{Bmatrix}. \tag{1.133}$$

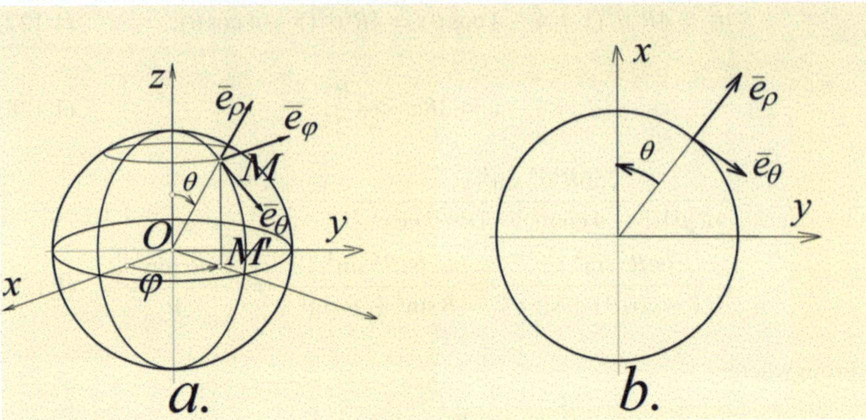

Fig. 1.36 Spherical coordinate system. **a** general presentation; **b** section through the sphere defined by a meridian

We have:

$$\{\dot{e}_\rho\} = \begin{Bmatrix} \cos\theta\cos\phi \\ \cos\theta\sin\phi \\ -\sin\theta \end{Bmatrix}\dot{\theta} + \begin{Bmatrix} -\sin\theta\sin\phi \\ \sin\theta\cos\phi \\ 0 \end{Bmatrix}\dot{\phi} = \dot{\theta}\{e_\theta\} + \dot{\phi}\sin\theta\{e_\phi\}; \quad (1.134)$$

$$\{\dot{e}_\theta\} = -\begin{Bmatrix} \sin\theta\cos\phi \\ \sin\theta\sin\phi \\ \cos\theta \end{Bmatrix}\dot{\theta} + \begin{Bmatrix} \cos\theta\sin\phi \\ -\cos\theta\cos\phi \\ 0 \end{Bmatrix}\dot{\phi} = -\dot{\theta}\{e_\rho\} + \dot{\phi}\cos\theta\{e_\phi\}; \quad (1.135)$$

$$\{\dot{e}_\phi\} = \begin{Bmatrix} -\cos\phi \\ -\sin\phi \\ 0 \end{Bmatrix}\dot{\phi} = -\dot{\phi}\sin\theta\{e_\rho\} - \dot{\phi}\{e_\theta\}. \quad (1.136)$$

The position vector of the material point is:

$$\vec{r} = r\vec{e}_\rho. \quad (1.137)$$

The velocity is obtained obtain via the relation:

$$\vec{v} = \dot{\vec{r}} = \dot{r}\vec{e}_\rho + r\dot{\vec{e}}_\rho = \dot{r}\vec{e}_\rho + r\dot{\theta}\,\vec{e}_\theta + r\dot{\phi}\sin\theta\,\vec{e}_\kappa, \quad (1.138)$$

so it has the components:

$$v_\rho = \dot{r};\; v_\theta = r\dot{\theta};\; v_\phi = r\dot{\phi}\sin\theta, \quad (1.139)$$

and the value:

$$v = \sqrt{v_\rho^2 + v_\theta^2 + v_\phi^2} = \sqrt{\dot{r}^2 + r^2\dot{\theta}^2 + r^2\dot{\phi}^2 \sin^2\theta}. \tag{1.140}$$

By derivation the acceleration is obtained:

$$\begin{aligned}\vec{a} = \dot{\vec{v}} &= \ddot{r}\vec{e}_\rho + \dot{r}\dot{\vec{e}}_\rho + \dot{r}\dot{\theta}\,\vec{e}_\theta + r\ddot{\theta}\,\vec{e}_\theta + r\dot{\theta}\,\dot{\vec{e}}_\theta \\ &+ \dot{r}\dot{\phi}\,\sin\theta\,\vec{e}_\phi + r\ddot{\phi}\,\sin\theta\,\vec{e}_\phi + r\dot{\phi}\dot{\theta}\cos\theta\,\vec{e}_\phi + r\dot{\phi}\,\dot{\vec{e}}_\phi.\end{aligned} \tag{1.141}$$

with the components:

$$a_\rho = \ddot{r} - r\dot{\theta}^2 - r\dot{\phi}^2\sin^2\theta, \quad a_\theta = 2\dot{r}\dot{\theta} + r\ddot{\theta} - r\dot{\phi}^2\sin\theta\cos\theta$$

$$a_\phi = r\ddot{\phi}\sin\theta + 2\dot{r}\dot{\phi}\sin\theta + 2\dot{\phi}\dot{\theta}r\cos\theta = \frac{1}{r\sin\theta}\frac{d(r^2\dot{\phi}\sin^2\theta)}{dt}. \tag{1.142}$$

1.4 Particular Movements of the Material Point

1.4.1 Rectilinear and Uniform

Definition Rectilinear and uniform motion is a motion in which the trajectory is a straight line and the velocity is constant.

It is most convenient to analyze this movement in the Cartesian coordinate system, considering that the movement is along the *Ox* axis. So:

$$v_x = v_0; \quad v_y = 0; \quad v_z = 0, \tag{1.143}$$

where:

$$v_x = \dot{x} = v_0 = ct \quad \Rightarrow \quad x = C_1 t + C_2. \tag{1.144}$$

The initial conditions allow the determination of the integration constants. Knowing that at the initial moment we have:

$$t = 0; \quad x = x_0; \quad v = v_0, \tag{1.145}$$

it results:

$$v_{x|t=0} = v_0 = C_1; \quad x_{|t=0} = x_0 = C_2. \tag{1.146}$$

So:

$$x = x_0 + v_0 t. \tag{1.147}$$

The acceleration will be: $a = \dot{v} = \ddot{x} = 0$. Reciprocally, the motion for which the acceleration is zero is the rectilinear and uniform motion. So:

$$\ddot{x} = 0; \quad \ddot{y} = 0; \quad \ddot{z} = 0, \tag{1.148}$$

leads to

$$\dot{x} = C_1; \quad \dot{y} = C_2; \quad \dot{z} = C_3; \tag{1.149}$$

$$x = C_1 t + C_4; \quad y = C_2 t + C_5; \quad z = C_3 t + C_6. \tag{1.150}$$

Considering the initial conditions, namely, at time $t = 0$ we have:

$$v_x = v_{0x}; \quad v_y = v_{0y}; \quad v_z = v_{0z}. \tag{1.151}$$

$$x = x_0; \quad y = y_0; \quad z = z_0, \tag{1.152}$$

it results:

$$x = x_0 + v_{0x} t; \quad y = y_0 + v_{0y} t; \quad z = z_0 + v_{0z} t. \tag{1.153}$$

Eliminating the time between the three relations above, the trajectory equation (a straight line) is obtained:

$$\frac{x - x_0}{v_{0x}} = \frac{y - y_0}{v_{0y}} = \frac{z - z_0}{v_{0z}}, \tag{1.154}$$

and the velocity value is:

$$v^2 = v_{0x}^2 + v_{0y}^2 + v_{0z}^2 = v_0^2 = ct. \tag{1.155}$$

1.4.2 Uniformly Varied Rectilinear Motion

Definition Uniformly varied rectilinear motion is a motion for which the trajectory is a straight line and the acceleration is constant.

For simplicity, consider the *Ox* axis in the direction of the mobile's trajectory. Then:

$$a_x = a; \quad a_y = 0; \quad a_z = 0; \tag{1.156}$$

$$v_x = a\,t + C_1; \quad v_y = C_2; \quad v_z = C_3; \tag{1.157}$$

$$x = \frac{a\,t^2}{2} + C_1\,t + C_4; \;\; y = C_2\,t + C_5; \;\; z = C_3\,t + C_6. \tag{1.158}$$

The initial conditions corresponding to the moment $t = 0$ are:

$$v_x = v_{0x}; \quad v_y = 0; \quad v_z = 0; \; x = x_0; \quad y = 0; \quad z = 0. \tag{1.159}$$

It results:

$$v_x = v = v_0 + a\,t; \quad v_y = 0; \;\; v_z = 0; \tag{1.160}$$

$$x = x_0 + v_{0x}\,t + \frac{a\,t^2}{2}; \qquad y = 0; \qquad z = 0. \tag{1.161}$$

If time is eliminated between the law of space and formula of velocity, we obtain Galilei's formula (Torricelli):

$$t = \frac{v - v_0}{a}; \qquad x = \frac{a}{2}\left(\frac{v - v_0}{a}\right)^2 + v_0\,\frac{v - v_0}{a} + x_0, \tag{1.162}$$

where from:

$$v^2 = v_0^2 + 2\,a\,(x - x_0). \tag{1.163}$$

When $v_O = 0$, $x_O = 0$ (the mobile leaves the origin with zero initial speed) we have:

$$v^2 = 2\,a\,x. \tag{1.164}$$

If acceleration and velocity have the same direction, the movement is called uniformly accelerated, and if they have opposite direction, it is called uniformly slowed down or delayed.

1.4.3 Circular Motion

Definition Circular motion is the movement of a material point on a fixed circle.

This movement can be studied in Cartesian (Fig. 1.37a), polar (Fig. 1.37b) or natural coordinates (Fig. 1.38).

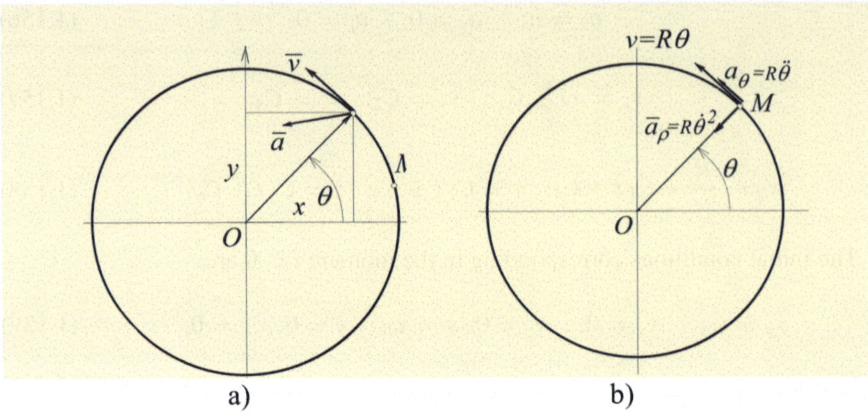

Fig. 1.37 **a** Cartesian coordinates; **b** Polar coordinates

Fig. 1.38 Natural coordinates

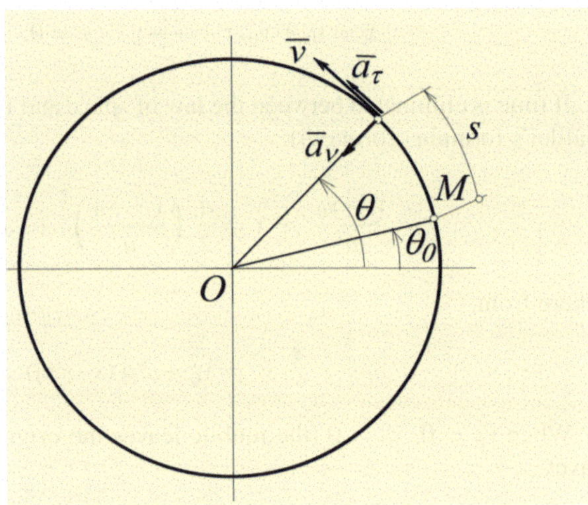

Thus, the Cartesian coordinates of the material point moving on the circle are:

$$x = R\cos\theta; \qquad y = R\sin\theta\ , \tag{1.165}$$

from where, by derivation, we obtain successively:

$$\dot{x} = -R\dot{\theta}\sin\theta; \qquad \dot{y} = R\dot{\theta}\cos\theta; \tag{1.166}$$

$$\ddot{x} = -R\ddot{\theta}\sin\theta - R\dot{\theta}^2\cos\theta; \qquad \ddot{y} = R\ddot{\theta}\cos\theta - R\dot{\theta}^2\sin\theta, \tag{1.167}$$

with:

$$v^2 = R^2\,\dot{\theta}^2 \;\Rightarrow\; v = R\left|\dot{\theta}\right| \quad and \quad a^2 = R^2\left(\ddot{\theta}^2 + \dot{\theta}^4\right). \tag{1.168}$$

If $\dot{\theta} = ct$, the motion is called uniform circular motion. Noting $\dot{\theta} = \omega_0$ results in $\theta = \omega_0\, t + \theta_0$. If $\ddot{\theta} > 0$ the movement is uniformly accelerated, and if $\ddot{\theta} < 0$ the movement is uniformly delayed. Noting $\dot{\theta} = \omega_0$ and $\ddot{\theta} = \varepsilon$ we obtain the law of space $\theta = \frac{\varepsilon\, t^2}{2} + \omega_0\, t + \theta_0$.

In polar coordinates (Fig. 1.37b) it is written:

$$\vec{r} = R \cdot \vec{e}_r; \tag{1.169}$$

$$\vec{v} = R\dot{\theta}\,\vec{e}_\theta; \qquad v = R\dot{\theta}; \tag{1.170}$$

$$\vec{a} = -R\dot{\theta}^2\,\vec{e}_r + R\ddot{\theta}\,\vec{e}_\theta; \qquad a = R\sqrt{\ddot{\theta}^2 + \dot{\theta}^4}; \tag{1.171}$$

with the components:

$$a_r = R\dot{\theta}^2; \qquad a_\theta = R\ddot{\theta}. \tag{1.172}$$

In natural coordinates (Fig. 1.38) we have:

$$s = AM = R\,(\theta - \theta_0); \tag{1.173}$$

$$v = \dot{s} = R\dot{\theta}; \qquad \vec{v} = \dot{s}\,\vec{\tau} = R\dot{\theta}\,\vec{\tau}; \tag{1.174}$$

$$a_\tau = \ddot{s} = R\ddot{\theta}; \qquad a_\nu = \frac{\dot{s}^2}{\rho} = R\dot{\theta}^2; \tag{1.175}$$

$$\vec{a} = R\ddot{\theta}\,\vec{\tau} + R\dot{\theta}^2\vec{\nu}; \tag{1.176}$$

$$tg\,\alpha = \frac{a_\tau}{a_\nu} = \frac{R\ddot{\theta}}{R\dot{\theta}^2} = \frac{\ddot{\theta}}{\dot{\theta}^2}. \tag{1.177}$$

Two images with mechanism in Middle Ages are presentes in Figs. 1.39 and 1.40.

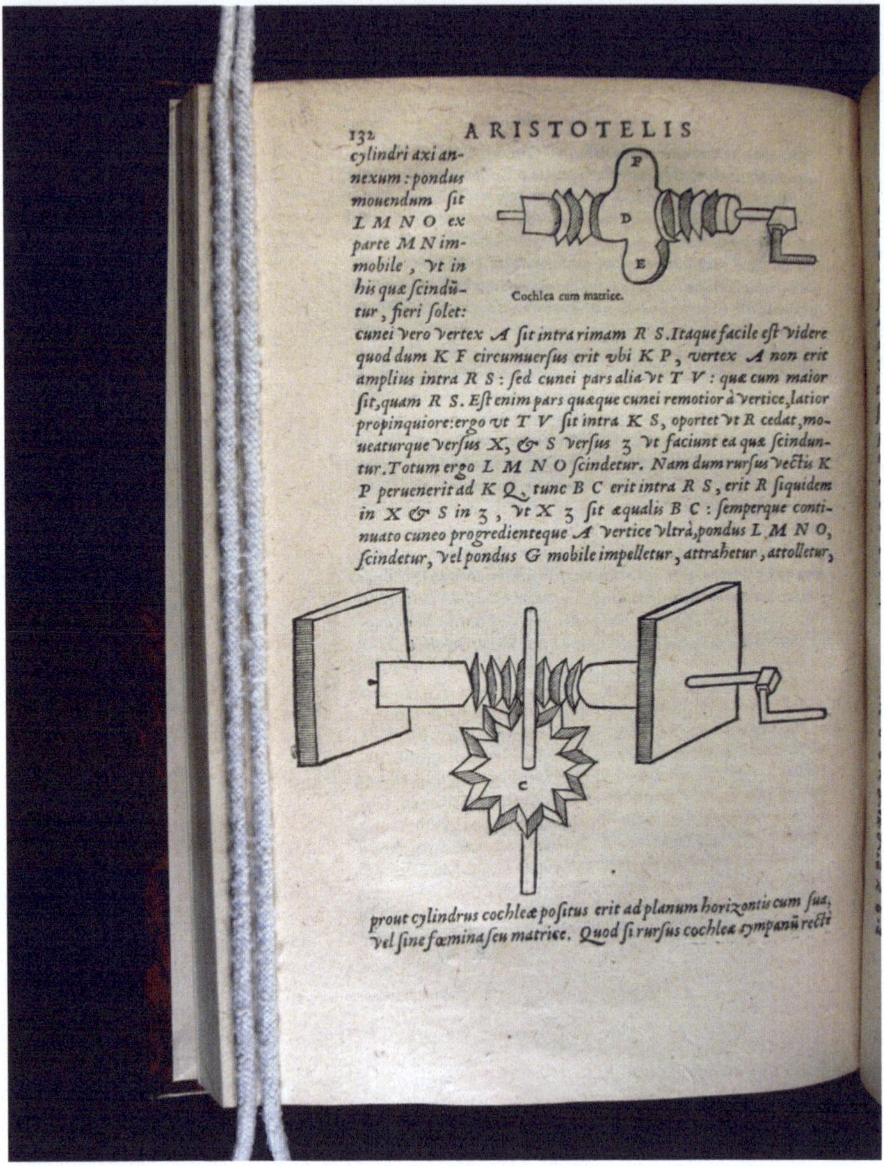

132 ARISTOTELIS

cylindri axi annexum: pondus mouendum sit L M N O ex parte M N immobile, vt in his quæ scindũtur, fieri solet:

Cochlea cum matrice.

cunei vero vertex A sit intra rimam R S. Itaque facile est videre quod dum K F circumuersus erit vbi K P, vertex A non erit amplius intra R S: sed cunei pars alia vt T V: quæ cum maior sit, quam R S. Est enim pars quæque cunei remotior à vertice, latior propinquiore: ergo vt T V sit intra K S, oportet vt R cedat, moueaturque versus X, & S versus ʒ vt faciunt ea quæ scinduntur. Totum ergo L M N O scindetur. Nam dum rursus vectis K P peruenerit ad K Q, tunc B C erit intra R S, erit R siquidem in X & S in ʒ, vt X ʒ sit æqualis B C: semperque continuato cuneo progredienteque A vertice vltrà, pondus L M N O, scindetur, vel pondus G mobile impelletur, attrahetur, attolletur,

prout cylindrus cochleæ positus erit ad planum horizontis cum sua, vel sine fœmina seu matrice. Quod si rursus cochleæ tympanũ rectè

Fig. 1.39 Page from Aristotelis Mechanica, 1599 [26]

Fig. 1.40 Systems of gears [27]

References

1. DEX Online (2026) Accessed 5 Feb 2026. Available at: https://dexonline.ro/
2. Brockhaus (1971) Lexikon der Naturwissenschaften und Technik, p 399
3. Encyclopedia Britannica (1980) Micropedia, vol 5, p 814
4. New York Public Library Digital Collections (2026) Accessed 5 Feb 2026. Available at: https://digitalcollections.nypl.org/collections/819ec2d0-c6bc-012f-7d85-58d385a7bc34
5. New York Public Library Digital Collections (2026) Accessed 5 Feb 2026. Available at:https://digitalcollections.nypl.org/items/825d16a0-c6bc-012f-d30c-58d385a7bc34?canvasIndex=0
6. Morin D (2008) Introduction to classical mechanics: with problems and solutions, 1st edn. Cambridge University Press
7. Taylor JR, Adam J (2020) Classical mechanics student solutions manual. University Science Books
8. Helliwell TM, Sahakian VV (2020) Modern classical mechanics. Cambridge University Press
9. Arnold VI, Weinstein A et al (1989) Mathematical methods of classical mechanics. Springer
10. Kibble TWB, Berkshire FH (2004) Classical mechanics, 5th edn. ICP
11. Reuleaux F, Ferguson ES (2013) The kinematics of machinery: outlines of a theory of machines. Dover Publications
12. Mare IC, Negrean I, Vlase S, Modrea A (2024) Gibbs-Appell equations used in finite element analysis of elastic materials with voids having a plane motion. Roman J Acoust Vibrat 21(1):83–90
13. Vlase S, Marin M, Elkhalfi A, Ailawalia P (2023) Mathematical model for dynamic analysis of internal combustion engines. J Comput Appl Mechan 54(4):607–622. https://jcamech.ut.ac.ir/article_94591.html
14. Cornell University Library Digital Collections (2026) Accessed 5 Feb 2026. Available at: https://digital.library.cornell.edu/catalog/ss:372659
15. Cornell University Library Digital Collections (2026) Accessed 5 Feb 2026. Available at: https://digital.library.cornell.edu/catalog/ss:372661

16. Cornell University Library Digital Collections (2026) Accessed 5 Feb 2026. Available at: https://digital.library.cornell.edu/catalog/ss:372662
17. MathCurve (2026) Cycloid spheric. Accessed 5 Feb 2026. Available at: https://mathcurve.com/courbes3d.gb/cycloidspheric/cycloidspheric.shtml
18. Cornell University Library Digital Collections (2026) Accessed 5 Feb 2026. Available at: https://digital.library.cornell.edu/catalog/ss:372666
19. Cornell University Library Digital Collections (2026) Accessed 5 Feb 2026. Available at: https://digital.library.cornell.edu/catalog/ss:372667
20. Stanciu MD, Vlase S, Marin M (2019) Vibration analysis of a guitar considered as a symmetrical mechanical system. Symmetry 11(6):727. https://www.mdpi.com/2073-8994/11/6/727
21. Vlase S, Teodorescu PP, Itu C, Scutaru ML (2013) Elasto-dynamics of a solid with general rigid motion using FEM model. Part II. Analysis of a double Cardan joint. Roman J Phys 58(7–8):882–892
22. Vlase S (1987) Elimination of Lagrangian multipliers. Mech Res Commun 14(1):17–22
23. Scutaru ML, Vlase S, Marin M (2023) Symmetrical mechanical system properties-based forced vibration analysis. J Comput Appl Mechan 54(4):501–514
24. Katouzian M, Vlase S, Marin M (2024) Elastic moduli for a rectangular fibers array arrangement in a two phases composite. J Comput Appl Mechan 55(3):538–551. https://www.mdpi.com/1996-1944/17/6/1334
25. Vlase S, Marin M, Elkhalfi A, Ailawalia P (2023) Mathematical model for dynamic analysis of internal combustion engines. J Comput Appl Mechan 54(4):607–622
26. Henri de Monantheuil (1599) Aristotelis mechanica
27. Paz EB, Ceccarelli M, Otero JE, Sanz JLM (n.d.) A brief illustrated history of machines and mechanisms, vol 10. Springer, Dordrecht

Chapter 2
Rigid Body Kinematics

2.1 The General Elements of the Pozition of the Rigid Body

2.1.1 *Defining the Position of a Rigid Body*

A rigid body has an infinity of material points. Knowing the motion of a rigid body comes back to knowing the motion of any of its points. The rigidity hypothesis requires that the distance between any two points remains constant. This hypothesis reduces the number of parameters to determine the movement of a rigid to only 6. Thus, if we have n points, the number of degrees of freedom is $3n$, and the minimum number of links to ensure the stiffness of the assembly is $3(n - 4) + 6$. Then the number of parameters needed to describe the movement of a rigid assembly of n points is the difference of the two numbers, namely 6. This value does not depend on the number of points that make up the rigid body [1–3]. To clarify the problem of defining the position of a rigid body, let us first consider the simpler case, i.e., a plane problem.

In the plane we can determine the position of a point belonging to a rigid body by attaching a local reference system Oxy. It is defined, in relation to the global reference system, by the coordinates of the origin $O(x_0, y_0)$ and by the angle θ made by the Ox axis with the Ox_1 axis. If we know the position of a point $M(x, y)$ in relation to the local reference system Oxy, rotated by the angle θ with respect to the global system, its position relative to the global reference system $O_1x_1y_1$ is determined with the relations:

$$\begin{Bmatrix} x_1 \\ y_1 \end{Bmatrix} = \begin{bmatrix} \cos\theta & -\sin\theta \\ \sin\theta & \cos\theta \end{bmatrix} \begin{Bmatrix} x \\ y \end{Bmatrix}, \tag{2.1}$$

known from elementary analytic geometry. It was noted:

S. Vlase et al., *Models in Kinematics for Engineers*,
https://doi.org/10.1007/978-3-032-28113-5_2

$$\{r_1\} = \{r\}_{Ox1y1} = \begin{Bmatrix} x_1 \\ y_1 \end{Bmatrix} = [R]\begin{Bmatrix} x \\ y \end{Bmatrix}; \tag{2.2}$$

$$\{r_0\} = \{r_0\}_{Ox1y1} = \begin{Bmatrix} x_0 \\ y_0 \end{Bmatrix}; \quad \{\rho\} = \{r\}_{Oxy} = \begin{Bmatrix} x \\ y \end{Bmatrix}; \tag{2.3}$$

$$[R] = \begin{bmatrix} \cos\theta & -\sin\theta \\ \sin\theta & \cos\theta \end{bmatrix}, \tag{2.4}$$

we have:

$$\{r_1\} = \{r_0\} + \{r\} = \{r_0\} + [R]\{\rho\}. \tag{2.5}$$

The vectors indexed with Ox_1y_1 indicate that they have the components expressed in the Ox_1y_1 coordinate system and those indexed with Oxy have the components expressed in that system. The notations tried to avoid the index to simplify the writing.

The same relations could be obtained by writing:

$$\vec{r}_1 = \vec{r}_0 + \vec{r} = x_{01}\vec{i}_1 + y_{01}\vec{j}_1 + x\,\vec{i} + y\vec{j} \quad ; \; \vec{r}_1 = x_1\vec{i}_1 + y_1\vec{j}_1. \tag{2.6}$$

And then taking into account the fact that (see Fig. 2.1b):

$$\begin{aligned} \vec{i} &= \cos\theta\,\vec{i}_1 + \sin\theta\,\vec{j}_1 \quad ; \\ \vec{j} &= -\sin\theta\,\vec{i}_1 + \cos\theta\,\vec{j}_1 \quad ; \end{aligned} \tag{2.7}$$

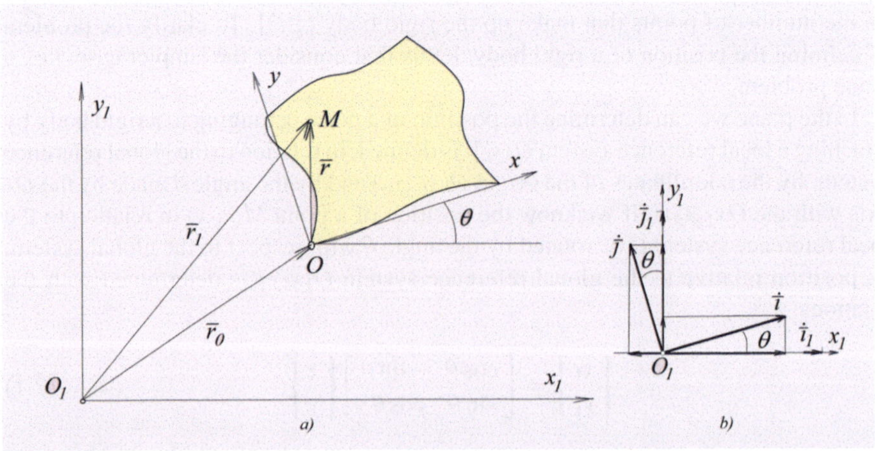

Fig. 2.1 Relating the rigid body to the local and global coordinate systems

by identifying the coefficients of $\vec{i}_1$ and $\vec{j}_1$ from the right and left members, the previous relationship is obtained:

$$\begin{aligned} x_1 &= x\cos\theta - y\sin\theta; \\ y_1 &= x\sin\theta + y\cos\theta, \end{aligned} \tag{2.8}$$

or:

$$\{r_1\} = \{r_0\} + [R]\{\rho\}, \tag{2.9}$$

where:

$$[R] = \begin{bmatrix} \cos\theta & -\sin\theta \\ \sin\theta & \cos\theta \end{bmatrix} = \big[\{i\}\ \{j\}\big], \tag{2.10}$$

represents the rotation matrix, which contains the unit vectors of the local reference system. It is easily verified that $[R]$ is an orthonormal matrix, so:

$$[R][R]^T = [R]^T[R] = [E], \tag{2.11}$$

and:

$$\det[R] = 1. \tag{2.12}$$

Analogously, let us consider a rigid body to which we attach a local coordinate system $Oxyz$ (Fig. 2.2). The position of the material point M in relation to the local system is known by its coordinates x, y, z. The position of point M in relation to the global coordinate system $Ox_1y_1z_1$ will be determined if the following are known:

- the position of the origin of the local reference system $Oxyz$;
- the orthogonal unit vectors $\vec{i}, \vec{j}$ and $\vec{k}$ of the three axes.

As an unit vector has three components, the knowledge of the three unit vectors would require the knowledge of nine scalar components. The conditions that $\vec{i}, \vec{j}$ and $\vec{k}$ represent the basis of an orthonormal system lead to six conditions:

$$\begin{aligned} &\vec{i}^2 = 1; \quad \vec{j}^2 = 1; \quad \vec{k}^2 = 1; \\ &\vec{i}\cdot\vec{j} = 0; \quad \vec{j}\cdot\vec{k} = 0; \quad \vec{k}\cdot\vec{i} = 0, \end{aligned} \tag{2.13}$$

so three scalar parameters are sufficient to characterize the position of the axes of the mobile reference system. If we also consider the three parameters that define the position of the origin O, it follows that the position of the local reference system, therefore also of the rigid body, is defined by six scalar parameters. The unit vectors $\vec{i}, \vec{j}$ and $\vec{k}$ are defined by the relations:

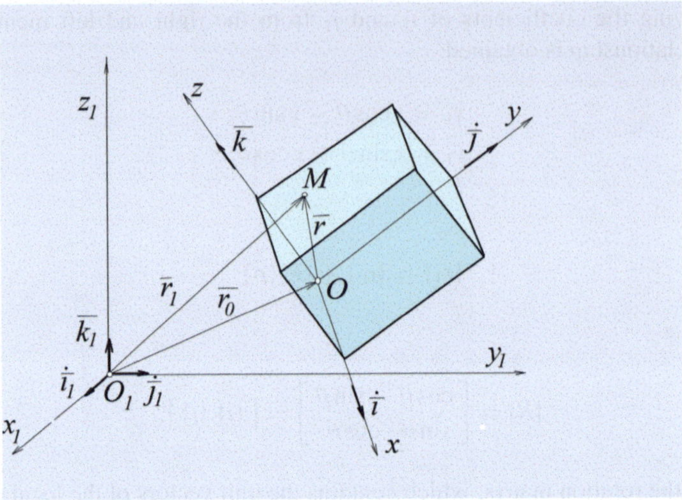

Fig. 2.2 Relating the rigid body to the local (mobile) and global (fixed) coordinate systems

$$\{i\} = \begin{Bmatrix} \cos(x, x_1) \\ \cos(x, y_1) \\ \cos(x, z_1) \end{Bmatrix}; \quad \{j\} = \begin{Bmatrix} \cos(y, x_1) \\ \cos(y, y_1) \\ \cos(y, z_1) \end{Bmatrix}; \quad \{k\} = \begin{Bmatrix} \cos(z, x_1) \\ \cos(z, y_1) \\ \cos(z, z_1) \end{Bmatrix}; \tag{2.14}$$

where $\cos(x, x_1)$ represents the cosine made by the Ox axis with the O_1x_1 axis (the projection of the unit vector $\vec{i}$ on the O_1x_1 axis), etc. We have the relationship:

$$\{r_1\} = \{r_0\} + \{r\}, \tag{2.15}$$

so:

$$x_1\vec{i}_1 + y_1\vec{j}_1 + z_1\vec{k}_1 = x_0\vec{i}_1 + y_0\vec{j}_1 + z_0\vec{k}_1 + x\,\vec{i} + y\,\vec{j} + z\,\vec{k}, \tag{2.16}$$

where by replacing $\vec{i}, \vec{j}$ and $\vec{k}$ and performing the calculations we get

$$\begin{Bmatrix} x_1 \\ y_1 \\ z_1 \end{Bmatrix} = \begin{Bmatrix} x_0 \\ y_0 \\ z_0 \end{Bmatrix} + \begin{bmatrix} \cos(x, x_1) & \cos(y, x_1) & \cos(z, x_1) \\ \cos(x, y_1) & \cos(y, x_1) & \cos(z, x_1) \\ \cos(x, z_1) & \cos(y, x_1) & \cos(z, x_1) \end{bmatrix} \begin{bmatrix} x \\ y \\ z \end{bmatrix}, \tag{2.17}$$

or:

$$\begin{Bmatrix} x_1 \\ y_1 \\ z_1 \end{Bmatrix} = \begin{Bmatrix} x_0 \\ y_0 \\ z_0 \end{Bmatrix} + \left[\,\{i\}\ \{j\}\ \{k\}\,\right] \begin{Bmatrix} x \\ y \\ z \end{Bmatrix} = \begin{Bmatrix} x_0 \\ y_0 \\ z_0 \end{Bmatrix} + [R] \begin{Bmatrix} x \\ y \\ z \end{Bmatrix}, \tag{2.18}$$

where it was noted:

$$[R] = \begin{bmatrix} \cos(x, x_1) \cos(y, x_1) \cos(z, x_1) \\ \cos(x, y_1) \cos(y, x_1) \cos(z, x_1) \\ \cos(x, z_1) \cos(y, x_1) \cos(z, x_1) \end{bmatrix} = \left[\{i\} \ \{j\} \ \{k\} \right]. \tag{2.19}$$

It is easy to verify that $[\mathbf{R}]$ is orthonormal, so verify the relation:

$$[R][R]^T = [R]^T[R] = [E], \tag{2.20}$$

a relation that imposes six scalar relations between the principal cosines. For ease of calculation, it was noted:

$$\{i\} = \begin{Bmatrix} \alpha_1 \\ \beta_1 \\ \gamma_1 \end{Bmatrix}; \quad \{j\} = \begin{Bmatrix} \alpha_2 \\ \beta_2 \\ \gamma_2 \end{Bmatrix}; \quad \{k\} = \begin{Bmatrix} \alpha_3 \\ \beta_3 \\ \gamma_3 \end{Bmatrix}. \tag{2.21}$$

The following scalar relations are obtained:

$$\begin{aligned} &\alpha_1^2 + \beta_1^2 + \gamma_1^2 = 1; \qquad \alpha_1\alpha_2 + \beta_1\beta_2 + \gamma_1\gamma_2 = 0 \\ &\alpha_2^2 + \beta_2^2 + \gamma_2^2 = 1; \qquad \alpha_2\alpha_3 + \beta_2\beta_3 + \gamma_2\gamma_3 = 0. \\ &\alpha_3^2 + \beta_3^2 + \gamma_3^2 = 1; \qquad \alpha_3\alpha_1 + \beta_3\beta_1 + \gamma_3\gamma_1 = 0 \end{aligned} \tag{2.22}$$

Of the nine cosine directors that appear in the rotation matrix $[\mathbf{R}]$ only three can be elected independent. In practice, the director cosines are rarely used as independent parameters, choosing three other parameters, for example Euler's angles, or Cardan's (Tait-Bryan) angles, easier to use in applications. The principal cosines will be expressed in terms of these independent parameters.

2.1.2 Finite Displacements. Infinitesimal Displacements

The displacement from an initial position to the final one is done through a sequence of movements. The relationship $\{r_1\} = \{r_0\} + [R]\{\rho\}$ characterizes only the initial and final position of the body without giving information about the path (intermediate positions) in which the movement was made.

In the case of moving in the plane (Fig. 2.3a) the vector $\{\mathrm{r}_0\}$ characterizes the translation, and the matrix $[R]$ contains the finite rotation angle θ by means of the sine and cosine functions. They define the finite displacement of the body in the plane. To reach the final position, the body passes through a sequence of intermediate positions (Fig. 2.3b), through infinitesimal displacements. The totality of the infinitesimal displacements gives the finite displacement.

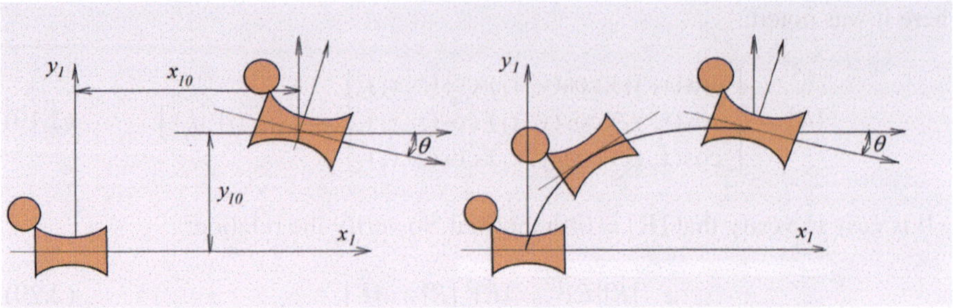

a) Initial and final position *b) Intermediate position*

Fig. 2.3 Plane paralel motion

In the case of a movement in space, the problem is posed in an analogous way. The finite displacement is characterized by the position vector $\{r_0\}$ of the origin of the mobile reference system and by the finite rotation angle of the rigid body around a straight line, to arrive with the axes from the initial position to the final one (Fig. 2.4).

We assume that the finite angle rotation φ was made around an axis that we will determine. During the rotation the axis remains unchanged, so if $\{u\}$ is the direction of the axis then:

$$[R]\{u\} = \{u\}. \tag{2.23}$$

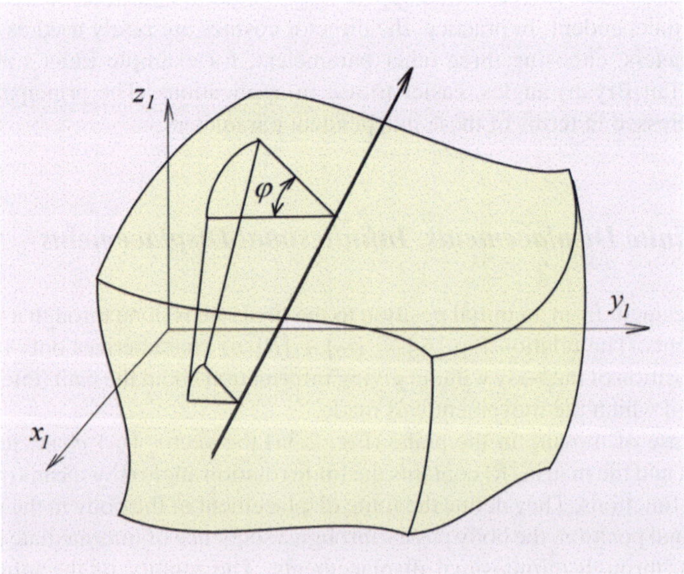

Fig. 2.4 Rotation of the rigid body around the instantaneous axis of rotation and translation

The following result is given, without proof: the matrix $[R]$ has the eigenvalues $1, e^{i\varphi}, e^{-i\varphi}$, where φ represents the finite rotation angle of the body around the axis determined by the versor $\{u\}$. In this case, the previous relation is immediately justified, as being the solution $\lambda = 1$ of the eigenvalue problem: $[\boldsymbol{R}]\ \{u\} = \lambda\{u\}$. To find the angle of rotation φ we take into account that the matrix $[\boldsymbol{R}]$ has the canonical form:

$$[R] = \begin{bmatrix} 1 & 0 & 0 \\ 0 & e^{i\phi} & 0 \\ 0 & 0 & e^{-i\phi} \end{bmatrix}. \tag{2.24}$$

The first invariant of the matrix is:

$$\mathrm{tr}[R] = 1 + e^{i\phi} + e^{-i\phi} = 1 + \cos\phi + i\sin\phi + \cos\phi - i\sin\phi = 1 + 2\cos\phi, \tag{2.25}$$

where we have:

$$\cos\phi = \frac{\mathrm{tr}[R]}{2}. \tag{2.26}$$

The trace of the matrix $[\boldsymbol{R}]$ can also be calculated with the relation:

$$\mathrm{tr}[R] = \alpha_{11} + \alpha_{22} + \alpha_{33}, \tag{2.27}$$

where from:

$$\cos\phi = \frac{\alpha_{11} + \alpha_{22} + \alpha_{33} - 1}{2}. \tag{2.28}$$

The relation $([R] - [E])\,\{u\} = 0$ gives us the direction of the axis of rotation. Consequently, to get from the initial position to the final one, a rigid body executes a vector translation movement $\{r_0\}$ and a rotation movement of angle φ around the vertical axis $\{u\}$.

2.1.3 Angular Velocity

It is considered a rigid body in general motion. The local reference system $Oxyz$ is attached to the rigid body, and we study the movement against the global reference system $O_1x_1y_1z_1$. We propose to determine the expression of the velocity of a point of the rigid body, assuming that we know the elements that characterize the movement of the local reference system (Fig. 2.5).

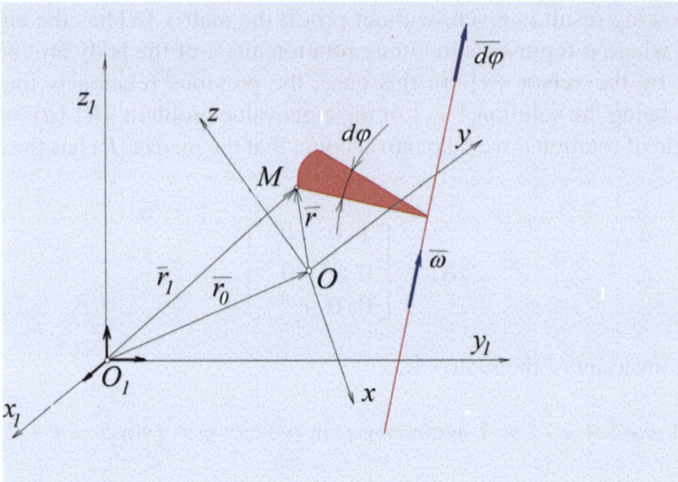

Fig. 2.5 The infinitesimal rotation of the rigid body

The components of the vector $\overrightarrow{OM}$ in the global coordinate system are denoted by x_1, y_1, z_1 and the components of the same vector in the local system by x, y, z. It is also noted $\vec{r} = \overrightarrow{OM}$ in the global system and $\vec{\rho} = \overrightarrow{OM}$ in the local system. The position vector $\vec{r}_M$ can be expressed by:

$$\{r_M\} = \{r_0\} + \{r\}; \quad (\vec{r}_M = \vec{r}_0 + \vec{r}). \tag{2.29}$$

If $[R]$ is the rotation matrix that makes the transition from the components of a vector expressed in the local system to the components of the same vector expressed in the global system, it can be written:

$$\{r_M\} = \{r_0\} + [R]\{\rho\}. \tag{2.30}$$

The components of the vector $\{\rho\}$ remain unchanged in time, the local system participating in the motion together with the rigid body. As a result, by derivation, the velocity of point M is obtained with the expression:

$$\{\dot{r}_M\} = \{\dot{r}_0\} + [\dot{R}]\{\rho\}. \tag{2.31}$$

Derivation of the relation:

$$[R][R]^{\mathrm{T}} = [E], \tag{2.32}$$

offers:

$$[\dot{R}][R]^T + [R][\dot{R}]^T = [0], \tag{2.33}$$

or, using the relation $([A][B])^{\mathrm{T}} = [B]^{\mathrm{T}}[A]^{\mathrm{T}}$, it results:

$$[\dot{R}][R]^T + \left([\dot{R}][R]^T\right)^T = [0]. \tag{2.34}$$

From here we conclude that the matrix $[\omega] = [\dot{R}][R]^T$ is skewymmetric because it verifies the relationship $[\omega] + [\omega]^{\mathrm{T}} = [0]$.

A skew matrix:

$$[\omega] = \begin{bmatrix} 0 & -\omega_z & \omega_y \\ \omega_z & 0 & -\omega_x \\ -\omega_y & \omega_x & 0 \end{bmatrix}, \tag{2.35}$$

can be used to represent a vector product. Next one can write:

$$\{\dot{r}_M\} = \{\dot{r}_0\} + [\dot{R}][R]^T[R]\{\rho\} = \{\dot{r}_0\} + [\omega][R]\{\rho\} = \{v_0\} + [\omega]\{r\}, \tag{2.36}$$

where: $\{v_0\}$ is the velocity of the origin of the moving reference frame. The product $[\omega]\{r\}$ represents the cross product between the vector $\{\omega\} = \begin{bmatrix} \omega_x & \omega_y & \omega_z \end{bmatrix}^{\mathrm{T}}$ associated with the operator $[\omega]$ and the vector $\{r\}$. It can also be written:

$$\vec{v}_M = \vec{v}_0 + \vec{\omega} \times \vec{r} \text{ (Euler's formula)}. \tag{2.37}$$

Writing the angular velocity by components, we get:

$$\begin{aligned} v_{Mx} &= v_{0x} + \omega_y z - \omega_z y; \\ v_{My} &= v_{0y} + \omega_z x - \omega_x z; \\ v_{Mz} &= v_{0z} + \omega_x y - \omega_y x\,. \end{aligned} \tag{2.38}$$

In the presented formulas all quantities are related to the local coordinate system. If it is related to the global system, the relations have the same form, but the quantities involved will be related to the global coordinate system.

The vector $\vec{\omega}$ is called the angular velocity, as it gives a representation of the rotational velocity of the mobile reference system, and the matrix $\vec{\omega}$ is called the angular velocity operator. The derived relation of the vector $\{r\}$ that moves together with the mobile reference system without changing its length and orientation with respect to the mobile reference system will also be retained:

$$\{\dot{r}\} = [\omega]\{r\}; \qquad (\dot{\vec{r}} = \vec{\omega} \times \vec{r}). \tag{2.39}$$

If we denote by $\{\Omega\}$ the angular velocity vector with the components related to the local reference system, we will have the relations:

$$\{\omega\} = [R]\{\Omega\}, \text{ respectivelly } \quad \{\Omega\} = [R]^{\mathrm{T}}\{\omega\}. \tag{2.40}$$

2.1.4 Properties of the Velocity Field

(a) The connection between the velocities of two points Q and P belonging to a rigid body is determined. We have:

$$\begin{aligned} \vec{v}_P &= \vec{v}_0 + \vec{\omega} \times \overrightarrow{OP}; \\ \vec{v}_Q &= \vec{v}_0 + \vec{\omega} \times \overrightarrow{OQ} \ . \end{aligned} \tag{2.41}$$

By subtraction we get:

$$\vec{v}_P - \vec{v}_Q = \vec{\omega} \times \left(\overrightarrow{OP} - \overrightarrow{OQ} \right) = \vec{\omega} \times \overrightarrow{PQ}, \tag{2.42}$$

or:

$$\vec{v}_P = \vec{v}_Q + \vec{\omega} \times \overrightarrow{PQ}\,. \tag{2.43}$$

(b) By projecting the previous relationship along the direction of the vector $\vec{\omega}$, we obtain:

$$\vec{v}_P\vec{\omega} = \vec{v}_Q\vec{\omega} + \left(\vec{\omega} \times \overrightarrow{PQ} \right) \vec{\omega} = \vec{v}_Q\vec{\omega}, \tag{2.44}$$

that is, the dot product of velocity and angular velocity $\vec{\omega}$ is constant: $\vec{v}\,\vec{\omega} = ct$. This shows us that, in general, there are no points of zero velocity. For certain particular cases of movement there are also such points. The relationship can be simplified as follows:

$$v_P\,\omega\,\cos\left(\vec{v}_P, \vec{\omega}\right) = v_Q\,\omega\,\cos\left(\vec{v}_Q, \vec{\omega}\right) \text{ or } v_P\,\cos\left(\vec{v}_P, \vec{\omega}\right) = v_Q\,\cos\left(\vec{v}_Q, \vec{\omega}\right). \tag{2.45}$$

So:

$$pr_{\vec{\omega}}\vec{v}_P = pr_{\vec{\omega}}\vec{v}_Q. \tag{2.46}$$

The projections of the velocities of different points of a rigid body in general motion, on the support of the angular velocity, are constant.

(c) By projecting the previous relation on the line PQ, we obtain:

$$\vec{v}_P\,\overrightarrow{QP} = \vec{v}_Q\,\overrightarrow{QP} + \vec{\omega} \times \left(\vec{\omega} \times \overrightarrow{QP}\right), \tag{2.47}$$

or:

$$\vec{v}_P\,\overrightarrow{QP} = \vec{v}_Q\,\overrightarrow{QP}\,. \tag{2.48}$$

$$v_P\,QP\,\cos\left(\vec{v}_P, \overrightarrow{QP}\right) = v_Q\,QP\,\cos\left(\vec{v}_Q, \overrightarrow{QP}\right), \tag{2.49}$$

or:

$$v_P\,\cos\left(\vec{v}_P, \overrightarrow{QP}\right) = v_Q\,\cos\left(\vec{v}_Q, \overrightarrow{QP}\right), \tag{2.50}$$

$$pr_{\overrightarrow{QP}}\vec{v}_P = pr_{\overrightarrow{QP}}\vec{v}_Q. \tag{2.51}$$

The Projections of Some Velocities of a Rigid Body in General Motion, on the Line that Joins Them, Are Equal. (Velocity Projections Theorem)

The rigidity of the adopted model is thus confirmed: the distance between any two points remains unchanged, since the components of the velocities along the line that joins them are equal.

(d) The extremities of the velocity vectors (drawn to scale) of collinear points belonging to a rigid body in general motion are collinear. (Theorem of the collinearity of the extremities of the velocity vectors).

If we consider the points A, B, C located on a straight line, we can write:

$$\overrightarrow{AB} = \lambda\,\overrightarrow{AC}, \tag{2.52}$$

or:

$$\vec{r}_A - \vec{r}_B = \lambda\,(\vec{r}_C - \vec{r}_A), \tag{2.53}$$

hence, by derivation:

$$\vec{v}_A - \vec{v}_B = \lambda\,(\vec{v}_C - \vec{v}_A). \tag{2.54}$$

In Fig. 2.6 it was considered: $AA' = k\,v_A$, $BB' = k\,v_B$, $CC' = k\,v_C$.

Multiplying the relation that connects the velocities by k and adding it to the one that connects the positions, we get:

$$(\vec{r}_B + k\,\vec{v}_B) - (\vec{r}_A + k\,\vec{v}_A) = \lambda\left[(\vec{r}_C + k\,\vec{v}_C) - (\vec{r}_A + k\,\vec{v}_A)\right], \tag{2.55}$$

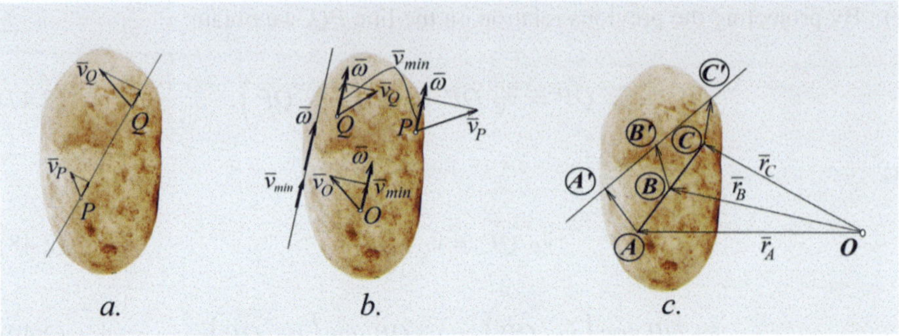

Fig. 2.6 Properties of the velocity field. **a** Velocity Projections Theorem; **b** The projections on the support of the angular velocity; **c** Theorem of the collinearity of the extremities of the velocity vectors

or:

$$\left(\vec{r}_B + \overrightarrow{BB'}\right) - \left(\vec{r}_A + \overrightarrow{AA'}\right) = \lambda\left[\left(\vec{r}_C + \overrightarrow{CC'}\right) - \left(\vec{r}_A + \overrightarrow{AA'}\right)\right], \tag{2.56}$$

or:

$$\overrightarrow{OB'} - \overrightarrow{OA'} = \lambda\left(\overrightarrow{OC'} - \overrightarrow{OA'}\right), \tag{2.57}$$

where from:

$$\overrightarrow{A'B'} = \lambda\,\overrightarrow{A'C'}, \tag{2.58}$$

which proves the collinearity of the points A', B', C'.

2.1.5 Angular Acceleration

By deriving the velocity, the expression of the acceleration of a point M is obtained:

$$\{a_M\} = \{\dot{v}_M\} = \{\ddot{r}_0\} + [\dot{\omega}]\{r\} + [\omega]\{r\} = \{\ddot{r}_0\} + [\dot{\omega}]\{r\} + [\omega][\omega]\{r\}. \tag{2.59}$$

If we denote $[\varepsilon] = [\dot{\omega}]$, taking into account the written relation, we get:

$$\{a_M\} = \{a_0\} + [\varepsilon]\{r\} + [\omega][\omega]\{r\}, \tag{2.60}$$

or, vectorially:

$$\vec{a}_M = \vec{a}_0 + \vec{\varepsilon} \times \vec{r} + \vec{\omega} \times (\vec{\omega} \times \vec{r}), \tag{2.61}$$

an expression called *Rivals' formula*, where $\vec{a}_0$ represents the origin of the mobile reference system. The vector $\vec{\varepsilon}$ is called the angular acceleration. The acceleration components, expressed in the Cartesian coordinate system, will be:

$$\begin{aligned} a_x &= a_{0x} + \varepsilon_y z - \varepsilon_z y - x\left(\omega_y^2 + \omega_z^2\right) + y\omega_x\omega_y + z\omega_x\omega_z; \\ a_y &= a_{0y} + \varepsilon_z x - \varepsilon_x z - y\left(\omega_z^2 + \omega_x^2\right) + z\omega_y\omega_z + x\omega_y\omega_x; \\ a_z &= a_{0z} + \varepsilon_x y - \varepsilon_y x - z\left(\omega_x^2 + \omega_y^2\right) + x\omega_z\omega_x + y\omega_z\omega_y\,. \end{aligned} \tag{2.62}$$

2.1.6 *Instant Axis of Rotation and Translation*

(a) Instant axis of rotation and translation (IART)

The problem arises of determining the set of points of a rigid body that have the minimum speed. If *A* and *B* are two points of the rigid body, the connection between the velocities of the two points is given by the relation (Fig. 2.7):

$$\vec{v}_B = \vec{v}_A + \vec{\omega} \times \vec{r}. \tag{2.63}$$

Multiplying in a dot product with $\vec{\omega}$ gives:

$$\vec{v}_B\vec{\omega} = \vec{v}_A\vec{\omega}, \tag{2.64}$$

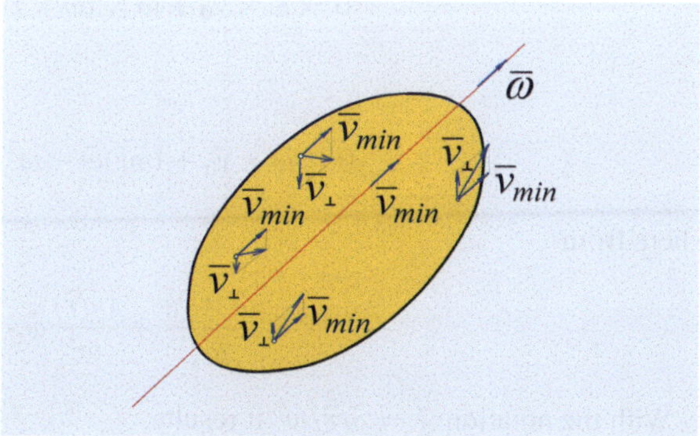

Fig. 2.7 Instantaneous axis of rotation and translation (IART)

or:

$$pr_{\vec{\omega}}\vec{v}_B = pr_{\vec{\omega}}\vec{v}_A, \tag{2.65}$$

that is, the projection on the direction of the angular velocity $\vec{\omega}$ of the velocity vector is constant for any point of the rigid body. It follows that the velocity vector has two components: one constant, following the direction of the angular velocity, v_ω and one perpendicular to it $v_\perp$: $\vec{v}_B = \vec{v}_\omega + \vec{v}_\perp$. The value of the speed is given by $v^2 = v_\omega^2 + v_\perp^2$ and is minimum at the moment when $v_\perp = 0$, since we saw previously that $\vec{v}_\omega = \overrightarrow{ct}$. In this case the velocity will be a vector parallel to the angular velocity $\vec{\omega}$:

$$\vec{v}_{\min} = \mu\,\vec{\omega}. \tag{2.66}$$

So we look for the position vector $\vec{r}$ as follows:

$$\vec{v}_{\min} = \mu\,\vec{\omega} = \vec{v}_0 + \vec{\omega} \times \vec{r}. \tag{2.67}$$

The previous relationship allows us, by multiplying by $\vec{\omega}$, to determine the constant μ:

$$\mu\,\omega^2 = \vec{v}_0\vec{\omega}; \qquad \mu = \frac{\vec{v}_0\omega}{\omega^2}, \tag{2.68}$$

from where the minimum velocity vector is:

$$\vec{v}_{\min} = \frac{\vec{v}_0\vec{\omega}}{\omega^2}. \tag{2.69}$$

If we multiply in a cross product with $\vec{\omega}$ we get:

$$0 = \vec{\omega} \times \vec{v}_0 + \vec{\omega} \times (\vec{\omega} \times \vec{r}), \tag{2.70}$$

or:

$$0 = \vec{\omega} \times \vec{v}_0 + (\vec{\omega}\vec{r})\vec{\omega} - \omega^2\vec{r}, \tag{2.71}$$

where from:

$$\vec{r} = \frac{\vec{\omega} \times \vec{v}_0}{\omega^2} + \frac{(\vec{\omega}\,\vec{r})}{\omega^2}\vec{\omega}. \tag{2.72}$$

With the notation: $\lambda = \vec{\omega}\,\vec{r}/\omega^2$ it results:

$$\vec{r} = \frac{\vec{\omega} \times \vec{v}_0}{\omega^2} + \lambda\vec{\omega}, \tag{2.73}$$

that is, the vector equation of a straight line. If we denote by $\vec{d} = \frac{\vec{\omega}\times\vec{v}_0}{\omega^2}$, a vector that represents the distance from point O to the found line, we obtain:

$$\vec{r} = \vec{d} + \lambda\,\vec{\omega}, \tag{2.74}$$

where from:

$$\begin{aligned} x &= d_x + \lambda\,\omega_x;\\ y &= d_y + \lambda\,\omega_y;\\ z &= d_z + \lambda\,\omega_z; \end{aligned} \tag{2.75}$$

(the equation of the line in parametric form). Eliminating λ gives:

$$\frac{x - x_0}{d_x} = \frac{y - y_0}{d_y} = \frac{z - z_0}{d_z}. \tag{2.76}$$

Another way to obtain the equation of the line would be to write the relation in terms of components:

$$\begin{aligned} \mu\,\omega_x &= v_{0x} + \omega_y z - \omega_z y;\\ \mu\,\omega_y &= v_{0y} + \omega_z x - \omega_x z;\\ \mu\,\omega_z &= v_{0z} + \omega_x y - \omega_y x\,. \end{aligned} \tag{2.77}$$

One of the three equations is a consequence of the other two. Indeed the determinant of the system

$$\begin{aligned} &-\omega_z y + \omega_y z = \mu\,\omega_x - v_{0x};\\ &\omega_z x - \omega_x z = \mu\,\omega_y - v_{0y};\\ &-\omega_y x + \omega_x y = \mu\,\omega_z - v_{0z}\,. \end{aligned} \tag{2.78}$$

is:

$$\Delta = \begin{bmatrix} 0 & -\omega_z & \omega_y\\ \omega_z & 0 & -\omega_x\\ -\omega_y & \omega_x & 0 \end{bmatrix} = 0 \tag{2.79}$$

and the characteristic determinant:

$$\Delta_C = \begin{bmatrix} \mu\,\omega_x - v_{0x} & -\omega_z & \omega_y\\ \mu\,\omega_y - v_{0y} & 0 & -\omega_x\\ \mu\,\omega_z - v_{0z} & \omega_x & 0 \end{bmatrix} = 0. \tag{2.80}$$

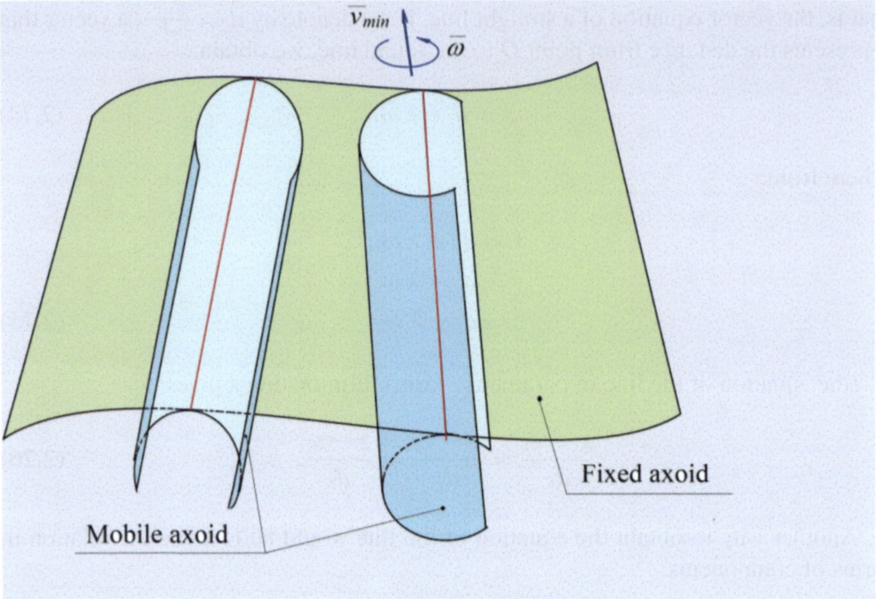

Fig. 2.8 Axoids

In conclusion: The set of points of the body that have the minimum velocity are organized on a straight line called the instantaneous axis of rotation and translation (IART). This axis is parallel to the angular velocity vector.

The motion of the rigid body can be considered instantaneously (Fig. 2.8) as a rotational motion around the IART, combined with a translation along this axis with the minimum speed $v_{\min} = (\vec{v}_0\vec{\omega})/\omega$ (hence the name of the instantaneous axis of rotation and translation).

(b) Velocity distribution around the instantaneous axis of rotation and translation

The following two properties will allow the representation of the velocity distribution around the instantaneous axis of rotation and translation.

The points belonging to a rigid in general motion and located on a parallel to the vector $\vec{\omega}$ have the same velocity vector.

Indeed, let us take the points *P* and *Q* on a line parallel to $\vec{\omega}$:

$$\vec{r}' = \vec{r} + \overrightarrow{PQ} = \vec{r} + \lambda\,\vec{\omega}$$

$$\begin{aligned}\vec{v}_Q &= \vec{v}_0 + \vec{\omega} \times \left(\vec{r} + \overrightarrow{PQ}\right)\\ &= \vec{v}_0 + \vec{\omega} \times (\vec{r} + \lambda\,\vec{\omega}) = \vec{v}_0 + \vec{\omega} \times \vec{r} = \vec{v}_P.\end{aligned} \tag{2.81}$$

The points belonging to a rigid body in general motion and located on the lateral surface of a cylinder having as axis the IART have equal velocity values.

We take point O on the IART. We have $\vec{v}_P = \vec{v}_0 + \vec{\omega} \times \vec{r}$. Let us take M on a line parallel to $\vec{\omega}$ that passes through P so that $OM \perp$ IART. Then:

$$\vec{v}_P = \vec{v}_0 + \vec{\omega} \times \left(\overrightarrow{OM} + \overrightarrow{MP} \right), \tag{2.82}$$

where from:

$$\vec{v}_P = \vec{v}_0 + \vec{\omega} \times \left(\vec{d} + \lambda \, \vec{\omega} \right) = \vec{v}_0 + \vec{\omega} \times \vec{d},$$

$$v_P^2 = v_0^2 + \omega^2 d^2, \tag{2.83}$$

from where, if $d = ct.$, $v_O = ct = v_{\min}$, $\omega = ct$ it follows that $v_P = ct$. These two properties allow the representation in Fig. 2.9.

c) The axoids of rigid motion

Definition. The surface generated by IART in its motion relative to the fixed coordinate system is called a fixed axoid, and the surface generated by IART in motion

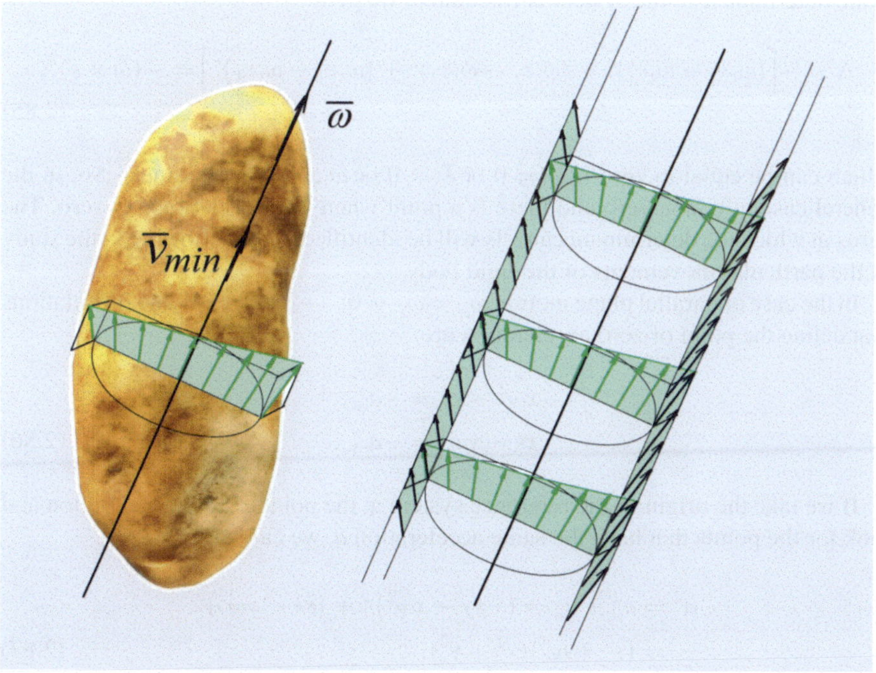

Fig. 2.9 Velocity distribution around IART

relative to the mobile coordinate system is called mobile axoid. During motion, the mobile axoid performs a rotation and sliding motion over the fixed axoid.

2.1.7 Pole of Accelerations

The problem arises of the existence of a point where the acceleration is zero. The condition $\vec{a} = 0$ leads to $a_x = a_y = a_z = 0$ and the following relations are obtained:

$$\begin{aligned} & -x\left(\omega_y^2+\omega_z^2\right)+y\left(\omega_x\omega_y-\varepsilon_z\right)+z\left(\omega_x\omega_z+\varepsilon_y\right)=-a_{0x}; \\ & x\left(\omega_x\omega_y+\varepsilon_z\right)-y\left(\omega_z^2+\omega_x^2\right)+z\left(\omega_y\omega_z-\varepsilon_x\right)=-a_{0y}; \\ & x\left(\omega_x\omega_z-\varepsilon_y\right)+y\left(\omega_y\omega_z+\varepsilon_x\right)-z\left(\omega_x^2+\omega_y^2\right)=-a_{0z}, \end{aligned} \tag{2.84}$$

which represents a linear system. The x, y, z solution of the system represents the point where the acceleration of the rigid body is zero (pole of acceleration). There may be cases of particular motions where the solution does not exist, so there are no points of zero acceleration or motions where the solution is undetermined (there are an infinity of points of zero acceleration, for example the points of a straight line). If the determinant of the system is calculated, we get:

$$\Delta=-\left[\left(\omega_y\varepsilon_z-\omega_z\varepsilon_y\right)^2+\left(\omega_z\varepsilon_x-\omega_x\varepsilon_z\right)^2+\left(\omega_x\varepsilon_y-\omega_y\varepsilon_x\right)^2\right]=-(\vec{\omega}\times\vec{\varepsilon})^2, \tag{2.85}$$

which can be equal to zero if $\vec{\omega} = 0$ or $\vec{\varepsilon} = 0$ or $\vec{\omega}$ and $\vec{\varepsilon}$ are collinear. So, in the general case, it is non-zero and there is a point where the acceleration is zero. The cases in which the determinant cancels will be identified in the following in the study of the particular movements of the rigid body.

In the case of parallel plane motion $\omega_x = \omega_y = 0;\quad \varepsilon_x = \varepsilon_y = 0$ and the relations that define the point of zero acceleration are:

$$\begin{aligned} -x\omega^2 - y\varepsilon &= -a_{0x}; \\ x\varepsilon - y\omega^2 &= -a_{0y}\ . \end{aligned} \tag{2.86}$$

If we take the origin of the reference system at the point of zero acceleration and look for the points that have the same acceleration a, we can write:

$$\begin{aligned} a^2 = a_x^2 + a_y^2 &= \left(-\varepsilon y - x\omega^2\right)^2 + \left(\varepsilon x - y\omega^2\right)^2 \\ &= \left(\varepsilon^2+\omega^4\right)(x^2+y^2), \end{aligned} \tag{2.87}$$

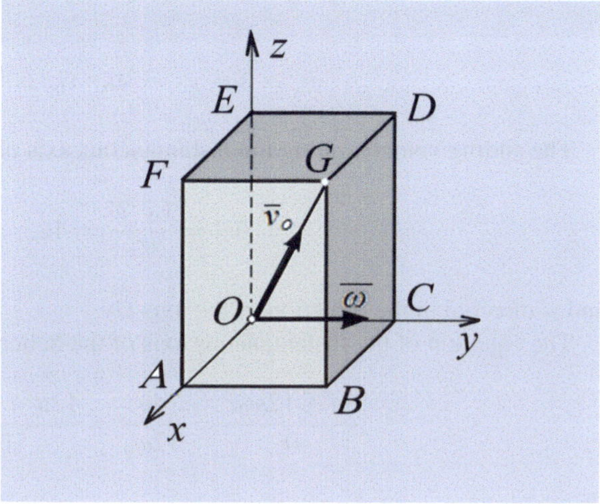

Fig. 2.10 Rotation of a rigid body arround an axis

so these points are found on a circle with the center at the point of zero acceleration. The distribution of accelerations is linear, along any straight line that passes through the pole of accelerations.

Examples ([4–7]).

E1. Given a parallelepiped $OABCDEFG$ of sides $OA = 3l$, $OC = 4l$, $OE = 12l$ that rotates with angular velocity $\omega = 12\omega_0$ around the edge OC during the motion, and the point O has a velocity $v_0 = 13$ m/s directed along the diagonal OG (Fig. 2.10). It is required to calculate the velocity of the point D and to establish the equation of the instantaneous axis of the helical motion.

Solution

Choose the system of axes in Fig. 2.10, with the origin in O and the axes directed along the edges of the parallelepiped. With respect to this reference system, we can write:

$$\overline{\omega} = 12\omega_0\overline{j}, \tag{2.88}$$

where from $\omega_x = \omega_z = 0$, $\omega_y = 12\omega_0$. The direction of unit vector OG is:

$$\overline{u}_{OG} = \frac{1}{3}\left(3\overline{i} + 4\overline{j} + 12\overline{k}\right), \tag{2.89}$$

from where it results:

$$\overline{v}_0 = 3u\,\overline{i} + 4u\,\overline{j} + 12u\,\overline{k}, \tag{2.90}$$

or:

$$v_{0x} = 3u;\ v_{0y} = 4u;\ v_{0z} = 12u. \tag{2.91}$$

The sliding velocity along the instantaneous axis of the helical motion is:

$$|\overline{v}_i| = \frac{\overline{v}_0 \cdot \overline{\omega}}{|\overline{\omega}|} = 4u, \tag{2.92}$$

and is directed in the direction of the axis Oy.

The equation of the instantaneous axis of the helical motion is:

$$\frac{3u + 12\omega_0 z}{0} = \frac{4u}{12\omega_0} = \frac{12u - 12\omega_0 x}{0}, \tag{2.93}$$

or:

$$\left.\begin{array}{l} x = \frac{u}{\omega_0} \\ z = -\frac{u}{4\omega_0} \end{array}\right\}, \tag{2.94}$$

i.e. a line parallel to the axis Oy, and therefore also to $\overline{\omega}$.

The velocity of the point D, using the general formula, is:

$$\overline{v}_D = \overline{v}_0 + \overline{\omega} \times \overline{OD} = 3u\,\overline{i} + 4u\,\overline{j} + 12u\,\overline{k} + \begin{vmatrix} \overline{i} & \overline{j} & \overline{k} \\ 0 & 12\omega_0 & 0 \\ 0 & 4l & 12l \end{vmatrix}$$
$$= (3u + 144\omega_0 l)\,\overline{i} + 4u\,\overline{j} + 12u\,\overline{k}. \tag{2.95}$$

The velocity of the point D can also be calculated as in a helical motion around the instantaneous axis of the helical motion. For this, the origin of the new reference system O' is chosen as the point where the instantaneous axis intersects the plane $OAFE$ and the axes parallel to the initial ones. With respect to the initial system, the point O' has the coordinates $x' = \frac{u}{\omega_0}$; $y' = 0$; $z' = -\frac{u}{4\omega_0}$. With respect to the new reference system, the point D has the coordinates $x'_D = x_D - x' = -\frac{u}{\omega_0}$; $y'_D = y_D - y' = 4l$; $z'_D = z_D - z' = 12l + \frac{u}{4\omega_0}$. Therefore, the velocity of the point D is:

$$\overline{v}_D = \overline{v}_i + \overline{\omega} \times \overline{O'D} = 4u\,\overline{j} + \begin{vmatrix} \overline{i} & \overline{j} & \overline{k} \\ 0 & 12\omega_0 & 0 \\ \frac{-u}{\omega_0} & 4l & 12l + \frac{u}{4\omega_0} \end{vmatrix}$$
$$= (3u + 144\omega_0 l)\,\overline{i} + 4u\,\overline{j} + 12u\,\overline{k}. \tag{2.96}$$

So, the same results.

The invariance of the velocity projection on the vector $\overline{\omega}$ support can be verified:

$$\overline{v}_D \frac{\overline{\omega}}{|\overline{\omega}|} = 4u \tag{2.97}$$

E2. Spatial rotations are noncommutative.

Solution

An example demonstrates the statement made in the title. Thus, for simplicity, let us call the axes *Ox, Oy, Oz* respectively axis *1*, axis *2* and axis *3*. Then, if we make two successive rotations of 90°, the first around axis *1* and the second around the new axis *2*, we obtain the body positioned as in Fig. 2.11, on the top right. If we make two rotations of 90° first around axis *2* and then around axis *1*, we obtain the body positioned as in Fig. 2.11, bottom right. The two positions of the body are, obviously, different.

Let us also do the calculation:

$$[R_\alpha] = \begin{bmatrix} 1 & 0 & 0 \\ 0 & \cos\frac{\pi}{2} & -\sin\frac{\pi}{2} \\ 0 & \sin\frac{\pi}{2} & \cos\frac{\pi}{2} \end{bmatrix} = \begin{bmatrix} 1 & 0 & 0 \\ 0 & 0 & -1 \\ 0 & 1 & 0 \end{bmatrix}, \tag{2.98}$$

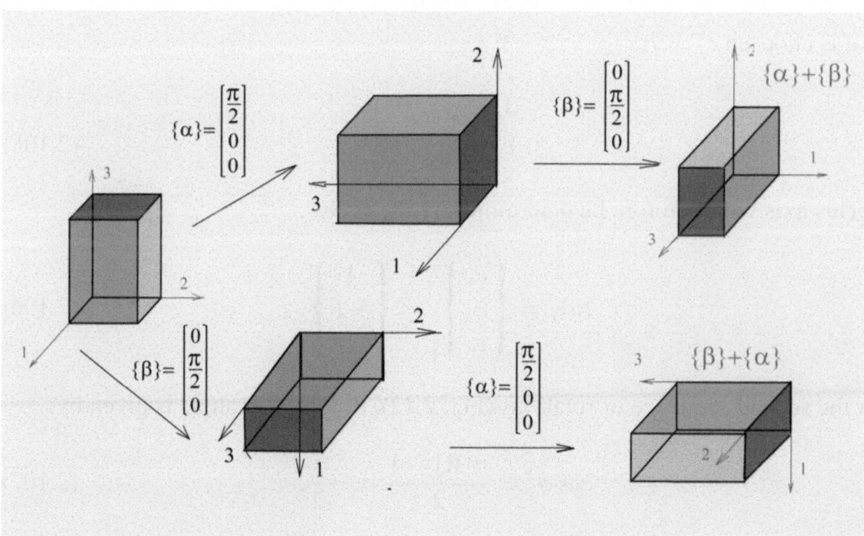

Fig. 2.11 Two succesive rotations

$$[R_\beta] = \begin{bmatrix} \cos\frac{\pi}{2} & 0 & -\sin\frac{\pi}{2} \\ 0 & 1 & 0 \\ \sin\frac{\pi}{2} & 0 & \cos\frac{\pi}{2} \end{bmatrix} = \begin{bmatrix} 0 & 0 & -1 \\ 0 & 1 & 0 \\ 1 & 0 & 0 \end{bmatrix}, \tag{2.99}$$

$$[R_{\alpha+\beta}] = [R_\beta][R_\alpha] = \begin{bmatrix} 0 & 0 & -1 \\ 0 & 1 & 0 \\ 1 & 0 & 0 \end{bmatrix}\begin{bmatrix} 1 & 0 & 0 \\ 0 & 0 & -1 \\ 0 & 1 & 0 \end{bmatrix} = \begin{bmatrix} 0 & -1 & 0 \\ 0 & 0 & -1 \\ 1 & 0 & 0 \end{bmatrix}, \tag{2.100}$$

$$[R_{\beta+\alpha}] = [R_\alpha][R_\beta] = \begin{bmatrix} 1 & 0 & 0 \\ 0 & 0 & -1 \\ 0 & 1 & 0 \end{bmatrix}\begin{bmatrix} 0 & 0 & -1 \\ 0 & 1 & 0 \\ 1 & 0 & 0 \end{bmatrix} = \begin{bmatrix} 0 & 0 & -1 \\ -1 & 0 & 0 \\ 0 & 1 & 0 \end{bmatrix}; \tag{2.101}$$

$$[R_{\alpha+\beta}] \neq [R_{\beta+\alpha}]. \tag{2.102}$$

For the first sequence of rotations, the rotation angle is given by:

$$\cos\phi_1 = \frac{\mathrm{tr}[\mathbf{R}] - 1}{2} = -\frac{1}{2}, \tag{2.103}$$

around the axis whose director parameters are determined by the equations:

$$\begin{bmatrix} -1 & -1 & 0 \\ 0 & -1 & -1 \\ 1 & 0 & -1 \end{bmatrix}\begin{Bmatrix} e_1 \\ e_2 \\ e_3 \end{Bmatrix}_1 = \begin{Bmatrix} 0 \\ 0 \\ 0 \end{Bmatrix}, \tag{2.104}$$

or, if is choosen $e_1 = 1$:

$$\begin{aligned} 1 + e_2 &= 0 \\ e_2 + e_3 &= 0. \end{aligned} \tag{2.105}$$

This axis will therefore be determined by the vector:

$$\{u\}_1 = \begin{Bmatrix} e_1 \\ e_2 \\ e_3 \end{Bmatrix}_1 = \begin{Bmatrix} 1 \\ -1 \\ 1 \end{Bmatrix}. \tag{2.106}$$

For the second sequence of rotations (Fig. 2.12), the rotation angle is given by:

$$\cos\phi_1 = \frac{\mathrm{tr}[\mathbf{R}] - 1}{2} = -\frac{1}{2}, \tag{2.107}$$

so the same as in the first case, but around the axis that has the director parameters determined by the equations:

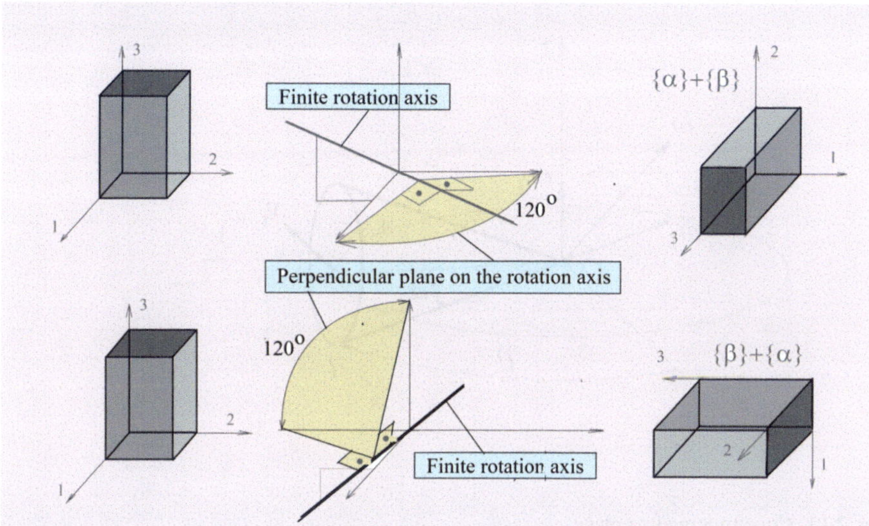

Fig. 2.12 Different order of plane rotations

$$\begin{bmatrix} -1 & 0 & -1 \\ -1 & -1 & 0 \\ 0 & 1 & -1 \end{bmatrix} \begin{Bmatrix} e_1 \\ e_2 \\ e_3 \end{Bmatrix}_2 = \begin{Bmatrix} 0 \\ 0 \\ 0 \end{Bmatrix}, \tag{2.108}$$

or, if is choosen $e_1 = 1$:

$$\begin{aligned} 1 + e_3 &= 0 \\ 1 + e_2 &= 0. \end{aligned} \tag{2.109}$$

This axis will therefore be determined by the vector:

$$\{u\}_2 = \begin{Bmatrix} e_1 \\ e_2 \\ e_3 \end{Bmatrix}_2 = \begin{Bmatrix} 1 \\ -1 \\ -1 \end{Bmatrix}. \tag{2.110}$$

The two vectors around which the finite rotation $\{u\}_1$ respectivelly $\{u\}_2$ takes place are therefore different, the rotation angle being in both cases 120°.

E3. A right circular cone of height h and radius r rolls without slipping on a plane, with the apex fixed at point O (Fig. 2.13). Calculate the angular velocity of the cone, when the velocity of the center of the base circle of the cone is known and constant ($v_C = ct.$).

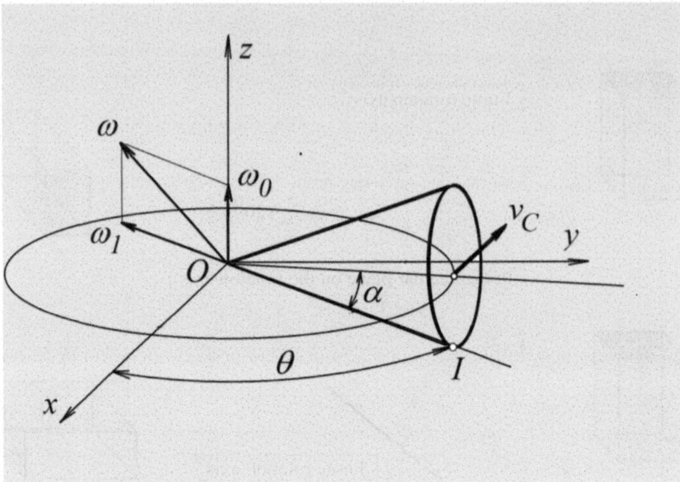

Fig. 2.13 Cone rotating in a plane

Solution

Let $\vec{\omega}_o$ be the angular velocity of rotation of the cone on the horizontal plane and $\vec{\omega}_1$ the angular velocity around the axis of the cone. The resulting angular velocity vector will be:

$$\vec{\omega} = \vec{\omega}_o + \vec{\omega}_1, \tag{2.111}$$

or

$$\vec{\omega} = -\omega_1 \cos\alpha\left(\cos\theta\,\vec{i} + \sin\theta\,\vec{j}\right) + (\omega_o - \omega_1 \sin\alpha)\vec{k}. \tag{2.112}$$

$$\vec{\omega} = -\omega_1 \cos\alpha\left(\cos\theta\,\vec{i} + \sin\theta\,\vec{j}\right) + (\omega_o - \omega_1 \sin\alpha)\vec{k}.$$

The velocity of the center of the base of the cone is:

$$\vec{v}_C = \omega_0 \cdot \vec{k} \times \vec{O}C, \tag{2.113}$$

where:

$$\vec{O}C = h\cos\alpha\left(\cos\theta\,\vec{i} + \sin\theta\,\vec{j}\right) + h\sin\alpha\,\vec{k}. \tag{2.114}$$

Substituting we obtain:

$$\vec{v}_C = h\omega_0 \cos\alpha\left(\cos\theta\,\vec{i} - \sin\theta\,\vec{j}\right), \tag{2.115}$$

with the modulus

$$\overline{v}_C = h\omega_0 \cos\alpha. \tag{2.116}$$

From here the angular velocity results

$$\omega_0 = \frac{v_C}{h\cos\alpha} = \frac{v_C\sqrt{r^2+h^2}}{h^2}. \tag{2.117}$$

As the cone rolls without slipping, OI is the instantaneous axis of rotation, so $v_I = 0$:

$$\vec{v}_I = \vec{\omega} \times \vec{OI} = \frac{h}{\cos\alpha}\begin{vmatrix} \vec{i} & \vec{j} & \vec{k} \\ \omega_x & \omega_y & \omega_z \\ \cos\theta & \sin\theta & 0 \end{vmatrix}$$
$$= \omega_z \sin\theta\,\vec{i} + \omega_z\cos\theta\,\vec{j} + \left(-\omega_y\cos\theta + \omega_x\sin\theta\right)\vec{k} = 0. \tag{2.118}$$

We therefore obtain:

$$\omega_z = 0 = \omega_0 - \omega_1 \sin\alpha, \tag{2.119}$$

and:

$$\omega_y = \omega_x tg\theta. \tag{2.120}$$

It results:

$$\omega_1 = \frac{\omega_0}{\sin\alpha} = \frac{v_C}{h\sin\alpha\cos\alpha} = v_C \cdot \frac{r^2+h^2}{h^2 r}. \tag{2.121}$$

References

1. Morin D (2008) Introduction to classical mechanics: with problems and solutions, 1st edn. Cambridge University Press
2. Taylor JR, Adam J (2020) Classical mechanics student solutions manual. University Science Books
3. Helliwell TM, Sahakian VV (2020) Modern classical mechanics. Cambridge University Press
4. Stanciu MD, Vlase S, Marin M (2019) Vibration analysis of a guitar considered as a symmetrical mechanical system. Symmetry 11(6):727. https://www.mdpi.com/2073-8994/11/6/727
5. Vlase S, Teodorescu PP, Itu C, Scutaru ML (2013) Elasto-dynamics of a solid with general rigid motion using FEM model. Part II. Analysis of a double Cardan joint. Roman J Phys 58(7–8):882–892

6. Scutaru ML, Vlase S, Marin M (2023) Symmetrical mechanical system properties-based forced vibration analysis. J Comput Appl Mechan 54(4):501–514. https://doi.org/10.22059/JCAMECH.2023.366509.890
7. Katouzian M, Vlase S, Marin M (2024) Elastic moduli for a rectangular fibers array arrangement in a two-phase composite. J Comput Appl Mechan 55(3):538–551. https://doi.org/10.22059/jcamech.2024.378143.1127

Chapter 3
Kinematics of Rigid Bodies. Particular Movements

3.1 Translational Motion

In Figs. 3.1 and 3.2 there are two examples of bodies that have a motion whose particularity will define the translational movement [1–3].

Definition. A solid has a translational motion if all its points describe equal and parallel trajectories.

This definition implies, as can be seen in Figs. 3.1 and 3.2, that any straight line of the solid remains parallel to itself. Sometimes this property is taken as the definition of translational motion.

Property. The translational movement is determined if the movement of a single point is known. Indeed, suppose that the trajectory (C) and the law of motion of a point M are known. The motion being translational, any straight line joining any two points of the body moves parallel to itself.

(a) The trajectory of a certain point P is deduced from the trajectory (C) through a translation determined by the vector $\overrightarrow{MP}$.
(b) The space traveled at time t by point P is equal to the space traveled in the same time interval by point M. We have $MM_1 = PP_1$ (the trajectories are parallel and equal).
(c) The velocity at time t of point P is equal to the velocity of point M. Since $MM_1 = PP_1$ we have:

$$\lim_{\Delta t \to 0} \frac{MM_1}{\Delta t} = \lim_{\Delta t \to 0} \frac{PP_1}{\Delta t}. \tag{3.1}$$

(d) The same property remains valid for accelerations. It follows that all points of the rigid body have, at a given moment, the same velocity and the same acceleration. This conclusion can be justified analytically.

S. Vlase et al., *Models in Kinematics for Engineers*,
https://doi.org/10.1007/978-3-032-28113-5_3

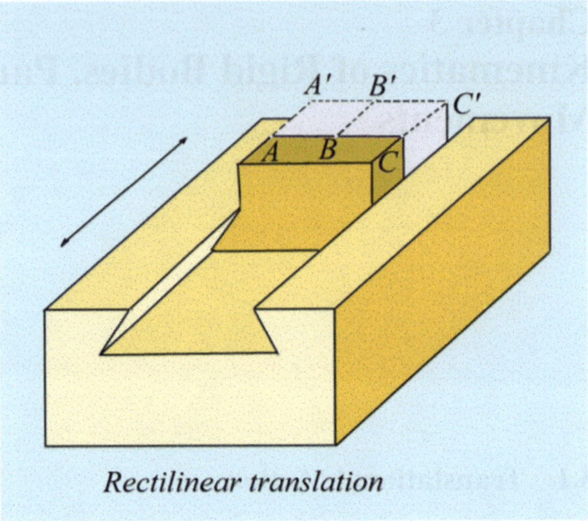

Fig. 3.1 Rectilinear translation

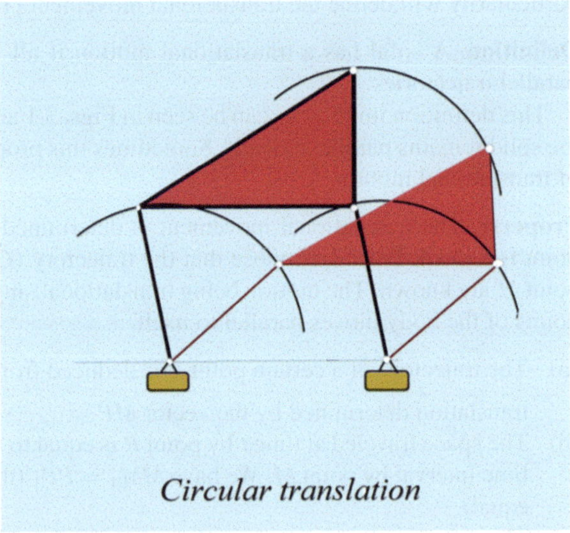

Fig. 3.2 Circular translation

From the definition of translational motion, it follows that any straight line always remains parallel to itself, so in particular the axes of the coordinate systems Ox, Oy, Oz (Fig. 3.3). The unit vectors of the three axes will not change their position during the movement, so:

$$\dot{\vec{i}} = 0\ ; \quad \dot{\vec{j}} = 0\ ; \quad \dot{\vec{k}} = 0 \tag{3.2}$$

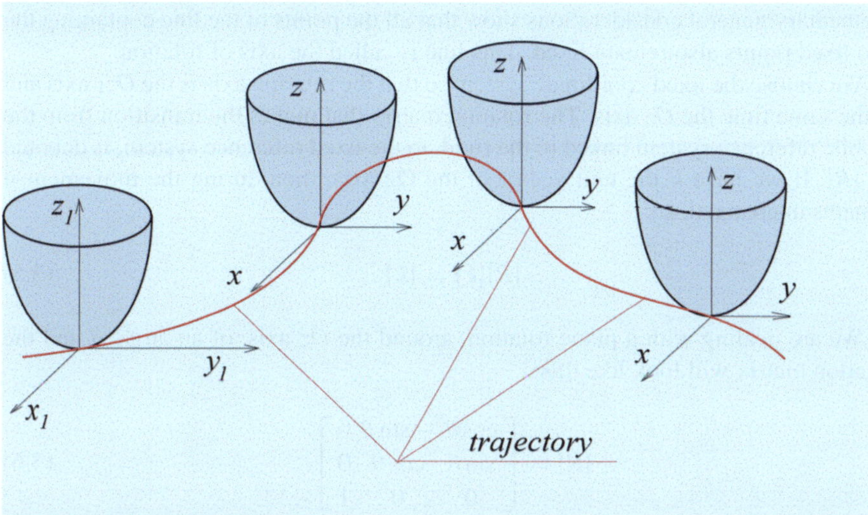

Fig. 3.3 Trajectory of general translations

immediately resulting in $\omega = 0$. The expression for the velocity of the current point M becomes:

$$\vec{v}_M = \vec{v}_0 + \vec{\omega} \times \vec{r} = \vec{v}_0; \tag{3.3}$$

and the acceleration:

$$\vec{a}_M = \vec{a}_0 + \vec{\varepsilon} \times \vec{r} + \vec{\omega} \times (\vec{\omega} \times \vec{r}) = \vec{a}_0. \tag{3.4}$$

According to the trajectory described by a point of the rigid body, the translations are divided into:

- rectilinear translations (Fig. 3.1) if the trajectory of a point is a straight line;
- curvilinear translations if the trajectory of a point is a certain curve in space (in particular if the trajectory is a circle we are dealing with a circular translation, Fig. 3.2).

3.2 Rotational Motion with a Fixed Axis

3.2.1 Expressions of Velocity and Acceleration

Definition. A rigid body has a rotational motion with a fixed axis if, during the entire movement, two points belonging to the rigid body remain fixed in space.

Elementary general considerations show that all the points of the line containing the two fixed points also remain fixed. This line is called the axis of rotation.

We choose the fixed coordinate system so that the rotation axis is the Oz_1 axis and at the same time the Oz axis. The rotation matrix that makes the transition from the mobile reference system linked to the rigid, to the fixed reference system, is denoted by $[R]$. If we have $\vec{k}$ the unit vector of the Oz axis, then during the movement it remains unchanged, so:

$$[R]\{k\} = \{k\}. \tag{3.5}$$

We are dealing with a plane rotation, around the Oz axis, of an angle θ and the rotation matrix will look like this:

$$[R] = \begin{bmatrix} \cos\theta & -\sin\theta & 0 \\ \sin\theta & \cos\theta & 0 \\ 0 & 0 & 1 \end{bmatrix}, \tag{3.6}$$

the first two columns representing the components of the vectors $\{i\}$ and $\{j\}$.

The angle θ is the angle made by the axes Ox_1 and Ox and the motion of the rigid body is defined by the function $\theta = \theta(t)$. The projection of M on the Oz_1 axis is denoted by M'. During the rotation, the distance OM' remains constant, the point M will move on a circle in a plane parallel to the plane Ox_1y_1. So the trajectories of all points are circles with the center on the axis Oz_1 (Fig. 3.4b). The position vector $\vec{r}$ has the expression:

$$\vec{r} = \overrightarrow{OM'} + \overrightarrow{MM'} = \overrightarrow{OM'} + \vec{d}, \tag{3.7}$$

where $\vec{d}$ is the vector $\overrightarrow{M'M}$ (distance vector of point M to the Oz axis). The angular velocity operator is:

$$\begin{aligned} [\omega] = [R][R]^T &= \dot{\theta} \begin{bmatrix} -\sin\theta & -\cos\theta & 0 \\ \cos\theta & -\sin\theta & 0 \\ 0 & 0 & 1 \end{bmatrix} \begin{bmatrix} \cos\theta & \sin\theta & 0 \\ -\sin\theta & \cos\theta & 0 \\ 0 & 0 & 1 \end{bmatrix} \\ &= \dot{\theta} \begin{bmatrix} 0 & -1 & 0 \\ 1 & 0 & 0 \\ 0 & 0 & 1 \end{bmatrix}, \end{aligned} \tag{3.8}$$

so the angular velocity vector is $[\omega] = \dot{\theta}\begin{bmatrix} 0 & 0 & 1 \end{bmatrix}^T$ i.e. it is a vector oriented along the Oz_1 axis ($\vec{\omega} = \omega\vec{k}$). By derivation, the angular acceleration vector $\varepsilon = \ddot{\theta}\begin{bmatrix} 0 & 0 & 1 \end{bmatrix}^T$ is obtained, therefore also oriented along the Oz_1 axis ($\vec{\varepsilon} = \dot{\vec{\omega}} = \varepsilon\vec{k}$). The speed of point M will be:

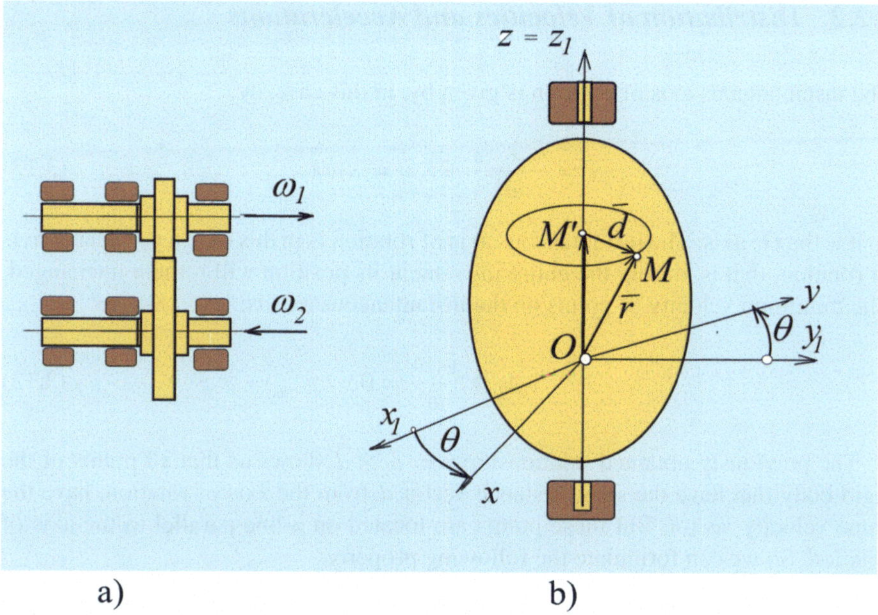

Fig. 3.4 Rotation around a fixed axis. **a** A real world application; **b** model

$$\vec{v} = \vec{v}_0 + \vec{\omega} \times \overrightarrow{OM} = \vec{v}_0 + \vec{\omega} \times \left(\overrightarrow{OM'} + \overrightarrow{M'M} \right)$$
$$= \vec{v}_0 + \vec{\omega} \times \overrightarrow{OM'} + \vec{\omega} \times \overrightarrow{M'M} = \vec{\omega} \times \vec{d} \tag{3.9}$$

so $\vec{v} = \vec{\omega} \times \vec{d}$ and $v = \omega\, d$ (because $\vec{\omega} \perp \vec{d}$).

The acceleration of a point M will have the expression:

$$\vec{a} = \vec{a}_0 + \vec{\varepsilon} \times \overrightarrow{OM} + \vec{\omega} \times \left(\vec{\omega} \times \overrightarrow{OM} \right)$$
$$= \vec{a}_0 + \vec{\varepsilon} \times \left(\overrightarrow{OM'} + \overrightarrow{M'M} \right) + \vec{\omega} \times \left(\vec{\omega} \times \vec{d} \right)$$
$$= \vec{\varepsilon} \times \vec{d} + \vec{\omega} \left(\vec{\omega} \vec{d} \right) - \omega^2 \vec{d} = \vec{\varepsilon} \times \vec{d} - \omega^2 \vec{d}. \tag{3.10}$$

The acceleration has two components, a tangential component $\vec{a}^{\tau} = \vec{\varepsilon} \times \vec{d}$ and a normal one $\vec{a}^{\nu} = -\omega^2 \vec{d}$. The acceleration value is given by:

$$a^2 = (a^{\tau})^2 + (a^{\nu})^2 = d^2(\varepsilon^2 + \omega^4) \Rightarrow a = d\sqrt{\varepsilon^2 + \omega^4}. \tag{3.11}$$

The formulas obtained for the circular movement of the material point can be found.

3.2.2 Distribution of Velocities and Accelerations

The instantaneous axis of rotation is given by, in this case, by:

$$\vec{r} = \frac{\vec{\omega} \times \vec{v}_0}{\omega^2} + \lambda\,\vec{\omega} = \lambda\,\omega\,\vec{k}, \tag{3.12}$$

so it is the Oz axis. The instantaneous axis of rotation is in this case a permanent axis of rotation, that is, during the entire movement its position will remain unchanged. The minimum velocity of points on the instantaneous axis is:

$$v_{\min} = \frac{\vec{v}_0\,\vec{\omega}}{\omega} = 0. \tag{3.13}$$

The previously obtained relationship $\vec{v} = \vec{\omega} \times \vec{d}$ shows us that all points of the rigid body that have the same distance vector $\vec{d}$ from the axis of rotation, have the same velocity vector. But these points are located on a line parallel to the axis of rotation. So we can formulate the following property:

- the set of points located on a line parallel to the axis of rotation (they have the same distance vector $\vec{d}$) have the same velocity vector.

  Another property resulting from this relationship is:

- the speeds are found in a plane perpendicular to the axis of rotation.

  The relation that gives the velocity modulus $v = \omega\,d$ allows us to formulate the second property of the velocity field:

- the set of points located at the same distance d from the axis have the same speed. Or: All points located on a right circular cylinder with the rotation axis as axis have the same velocity value.

  From this property derives:

- the velocity is directly proportional to the distance to the axis of rotation.

  These properties allow us to visualize the speed distribution as in Fig. 3.5.

  The relations previously obtained for accelerations: $\vec{a} = \vec{\varepsilon} \times \vec{d} - \omega^2\vec{d}$ and $a = d\sqrt{\varepsilon^2 + \omega^4}$ allow us to formulate, as in the case of speeds, the following properties:

- the set of points located on a line parallel to the axis of rotation (they have the same distance vector $\vec{d}$) have the same acceleration vector;
- the set of points located at the same distance d from the axis have the same acceleration value; or: All points located on a right circular cylinder, having the rotation axis as axis, have the same acceleration value;
- accelerations are directly proportional to the distance to the axis of rotation;
- accelerations are found in a plane perpendicular to the axis of rotation.

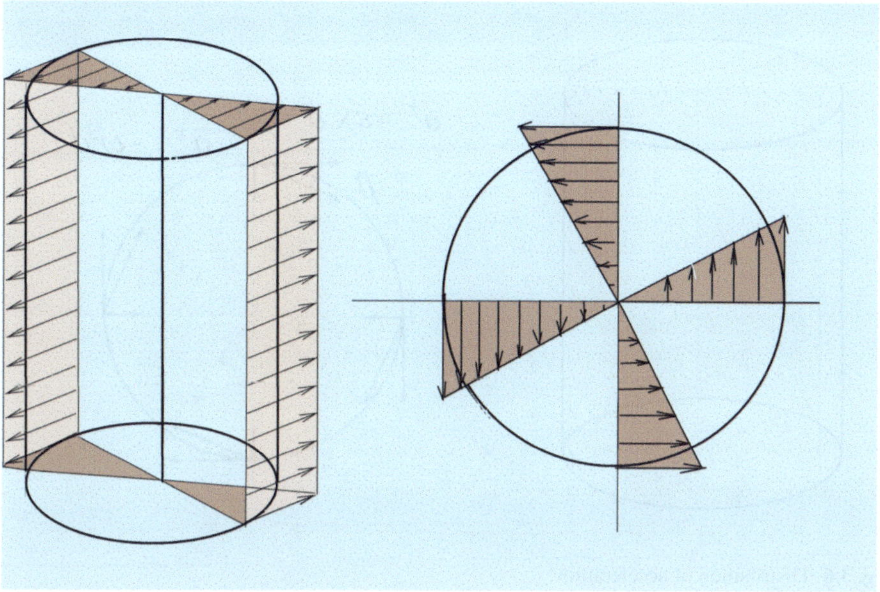

Fig. 3.5 Velocity distribution in rotational motion with a fixed axis

These properties allow us to visualize the distribution of accelerations as in Fig. 3.6. The angle made by the acceleration with the tangent to the trajectory is given by the relation:

$$ctg\beta = \frac{a^{\tau}}{a^{\nu}} = \frac{\varepsilon}{\omega^2}. \tag{3.14}$$

3.3 Helical (Rotational) Motion

3.3.1 Expressions of Velocity and Acceleration

Examples of bodies performing a movement that we will call helical are a drill during the drilling operation or a screw during the movement (Fig. 3.7).

Definition. A rigid body has a helical motion if there is a solid line which maintains the fixed support during the movement (Fig. 3.7).

The respective line is called the axis of the helical movement. To study this movement, the coordinate system is chosen so that the axis Oz and Oz_1 coincide with the axis of rotation. The body has a rotational movement around the axis and at the same time the origin of the mobile reference system has a translation along this

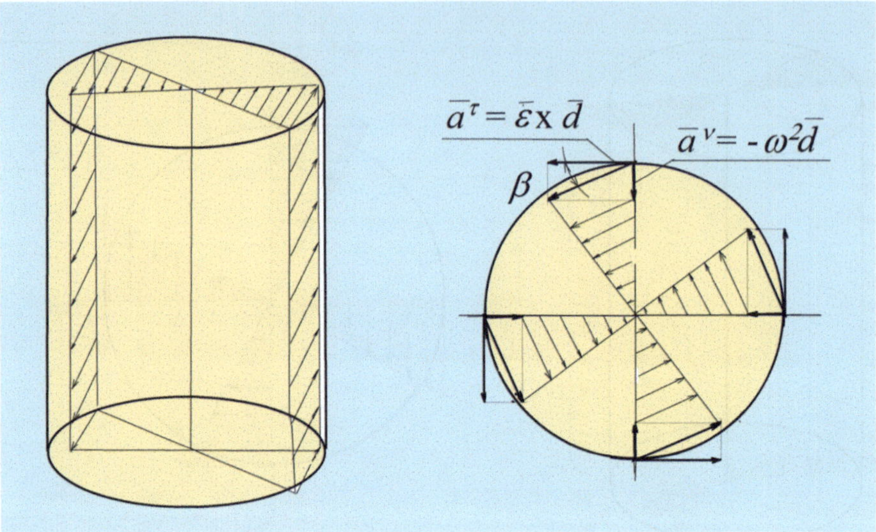

Fig. 3.6 Distribution of acceleration

axis according to a law $z_o = z_o(t)$. The rotational motion is characterized in the same way as in the case of the rotational motion with a fixed axis by the rotation angle $\theta = \theta\,(t)$ that expresses the rotation of the Ox axis with respect to the O_1x_1 axis.

By the same reasoning as in the rotational motion with a fixed axis, it will result that the angular velocity and angular acceleration of the body will be oriented along the O_1z_1 axis, so $\vec{\omega} = \omega\,\vec{k} = \dot{\theta}\,\vec{k}$ and $\vec{\varepsilon} = \varepsilon\,\vec{k} = \ddot{\theta}\,\vec{k}$ and the velocity and acceleration of the moving reference system will also be oriented along of this axis (Figs. 3.8 and 3.9):

$$\vec{v}_0 = \dot{z}_0\vec{k} = v_0\vec{k},\, \vec{a}_0 = \ddot{z}_0\vec{k} = a_0\vec{k}. \tag{3.15}$$

The velocity components result:

$$v_x = -y\omega;\ v_y = x\omega;\ v_z = v_0, \tag{3.16}$$

and of the acceleration:

$$a_x = -y\varepsilon - x\omega^2;\ a_y = x\varepsilon - y\omega^2;\ a_z = a_0. \tag{3.17}$$

If there is a linear relationship between the angle of rotation θ and the advance of the body along the z-axis, then we are dealing with a particular case, namely screw motion, which is particularly common in engineering (screws, presses, lathes).

a) Screw

b) Plastic cork

c) Drill bits

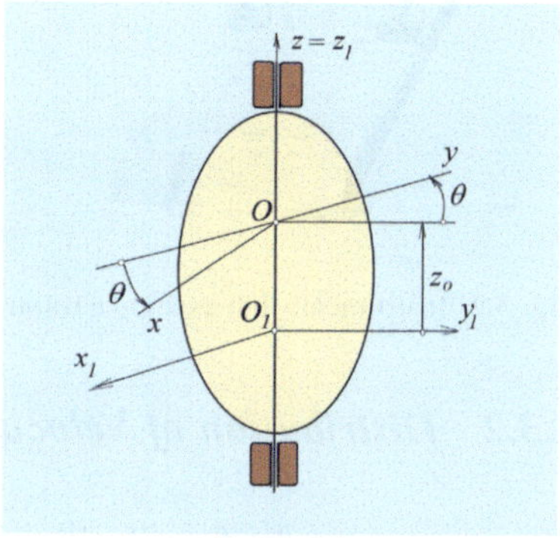

d) model

Fig. 3.7 Helical motion. Samples

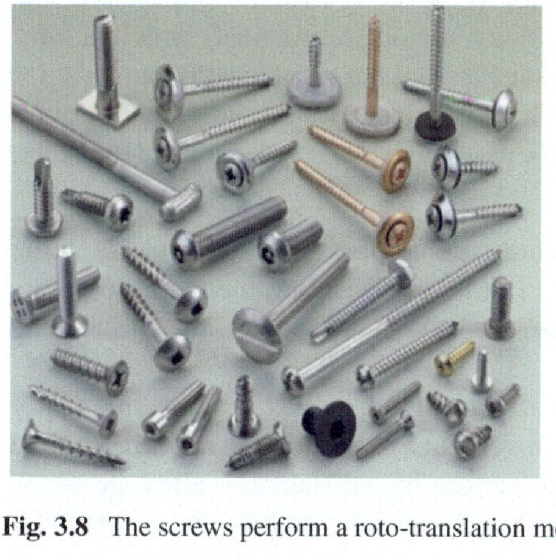

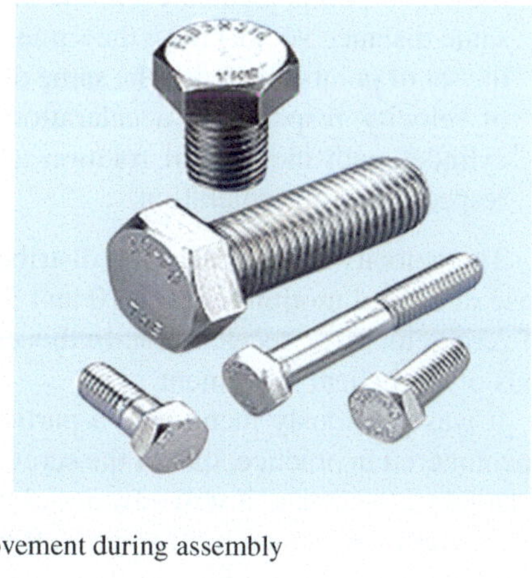

Fig. 3.8 The screws perform a roto-translation movement during assembly

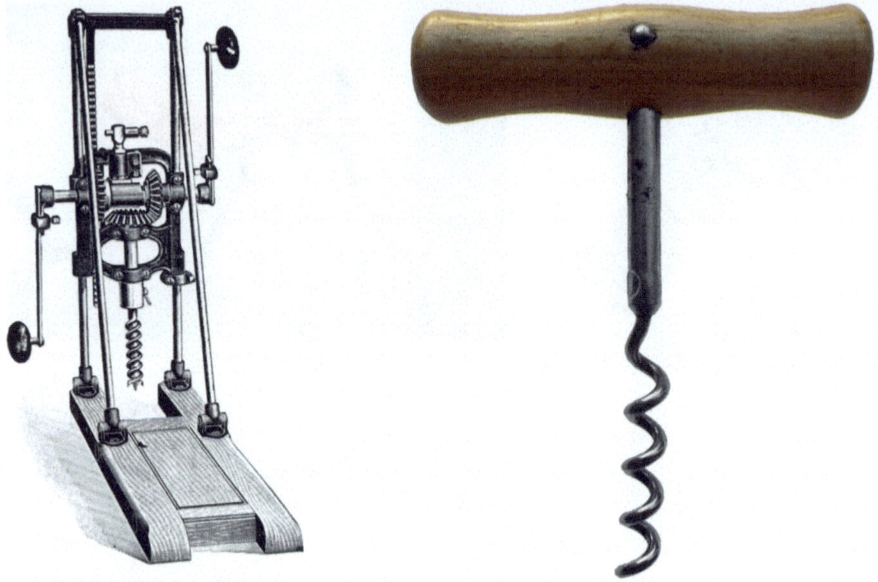

Fig. 3.9 Instruments that execute a rototranslation movement [4, 5]

3.3.2 Distribution of Velocities and Accelerations

The written formulas for velocities and accelerations allow, just like in the rotational motion with a fixed axis, to represent the distribution of velocities and accelerations in the helical motion. The following conclusions can be drawn:

- the set of points located on a line parallel to the axis of rotation (they have the same distance vector) have the same velocity vectors, respectively acceleration;
- the set of points located at the same distance d from the axis have the same values of velocity, respectively acceleration. Or: All points located on a right circular cylinder with the axis of rotation as its axis have the same value of velocity, respectively of acceleration.

The velocity and acceleration distributions can be represented in this way in the case of helical motion (Figs. 3.10 and 3.11). It is observed that there are no points of zero velocity or acceleration. In this case, the instantaneous axis of rotation is the axis of the helical movement.

It was previously mentioned a particular case of the helical motion, frequently encountered in practice, that of the screw movement. In this motion, at one complete rotation of the screw, it will advance along the axis by a distance p called the screw pitch. Between the angle of rotation θ and the displacement along the z axis there will be a connection which in this case is given by a linear relationship: $z = \frac{p}{2\pi}\theta$ (see the helix movement of the material point). By derivation it follows immediately: $v_0 = \frac{p}{2\pi}\omega$ and $a_0 = \frac{p}{2\pi}\varepsilon$.

Fig. 3.10 Velocities

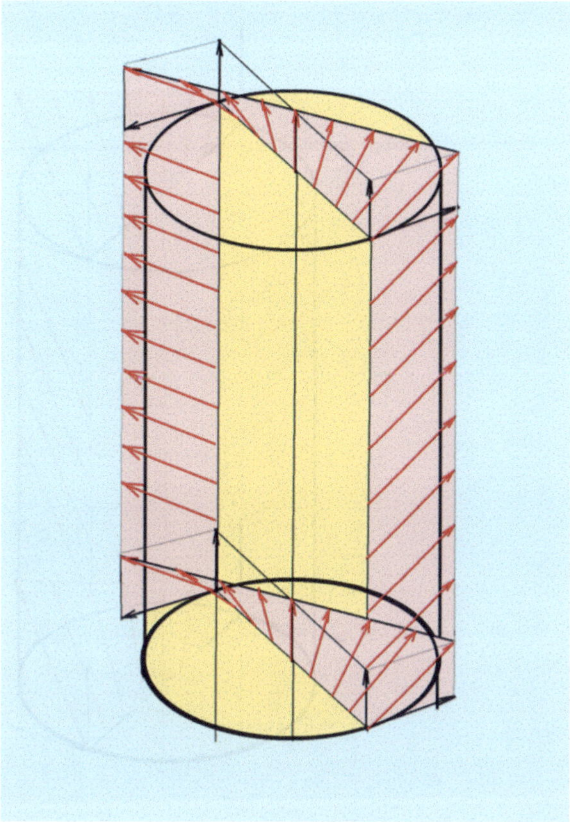

3.4 Plane-Parallel Motion

3.4.1 Expressions of Velocity and Acceleration

Figure 3.12 shows some examples that will characterize the plane-parallel movement.

Definition. A rigid body has plane-parallel motion if it has three non-collinear points that are contained throughout the motion in a fixed plane.

Elementary geometric considerations show that all points of the rigid body have trajectories contained in planes parallel to the fixed plane. We will show that it is sufficient to study the motion in a plane. Let us take the fixed plane to be the $O_1x_1y_1$ coordinate plane. Writing:

$$\{r\} = \{r_0\} + [R]\{\rho\}; \tag{3.18}$$

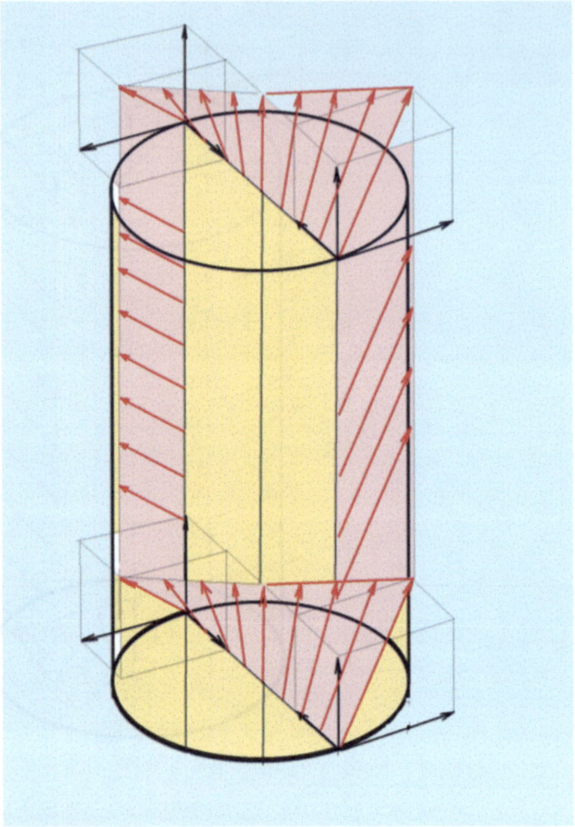

Fig. 3.11 Accelerations

a) translation

b) general plane motion

Fig. 3.12 Plane-parallel motion [6]

and taking into account the definition of motion, it will result that at all times the coordinate z = ct, so $\dot{z} = 0$. Then the matrix $[R]$ applied to the versor $\{k\}$ will not change it, so:

$$[R]\{\mathrm{k}\} = \{\mathrm{k}\}. \tag{3.19}$$

As with the rotational movement with a fixed axis, it follows that the $[\boldsymbol{R}]$ matrix has only five components, of which, taking into account the orthogonality relations, two remain independent. Then we will be able to note: $\alpha_{11} = \cos\theta$, $\alpha_{21} = \sin\theta$, $\alpha_{12} = -\sin\theta$, $\alpha_{22} = \cos\theta$. It results, according to Fig. 3.13:

$$\begin{Bmatrix} x_1 \\ y_1 \\ z_1 \end{Bmatrix} = \begin{Bmatrix} x_0 \\ y_0 \\ z_0 \end{Bmatrix} + \begin{bmatrix} \cos\theta & -\sin\theta & 0 \\ \sin\theta & \cos\theta & 0 \\ 0 & 0 & 1 \end{bmatrix} \begin{Bmatrix} x \\ y \\ z \end{Bmatrix}. \tag{3.20}$$

The z coordinates being constant in time, it follows that the parameters that will describe the movement will be x_0, y_0 and θ. It follows that the rigid body has three degrees of freedom. The trajectories will be defined by the variation of the three quantities over time. They will be plane curves. All points located on a line parallel to Oz have the same trajectory. It follows that in plane-parallel motion it is sufficient to study the motion in one plane (Fig. 3.14b). In the $O_1x_1y_1$ plane, the mobile reference system Oxy is chosen, which participates in the movement together with the rigid one. The angular speed of the rigid will be the speed of variation of the angle θ in

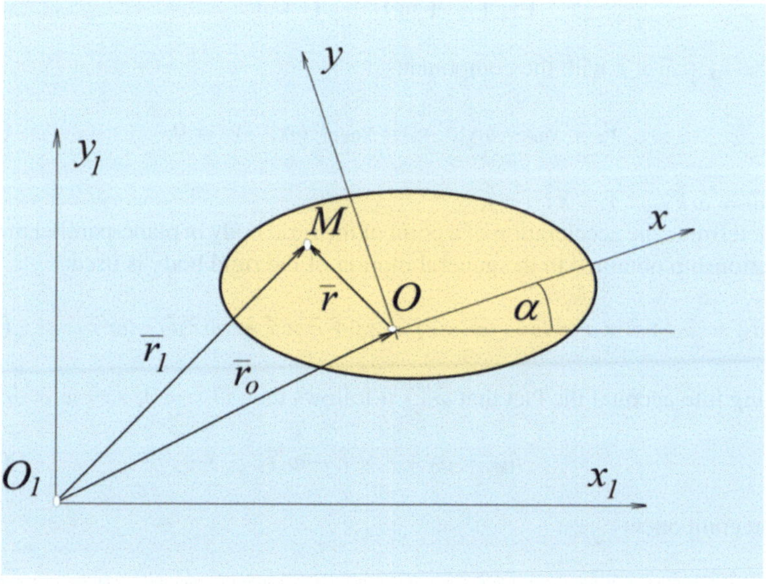

Fig. 3.13 Plan parallel motion of a rigid body

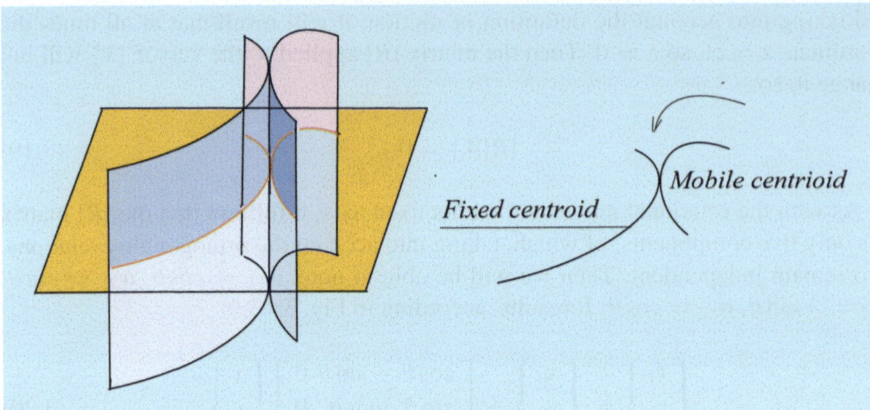

Fig. 3.14 The axoids (cylinders) and centroids (on plane)

relation to time. The previous relation can be written:

$$\begin{Bmatrix} x_1 \\ y_1 \end{Bmatrix} = \begin{Bmatrix} x_0 \\ y_0 \end{Bmatrix} + \begin{bmatrix} \cos\theta & -\sin\theta \\ \sin\theta & \cos\theta \end{bmatrix} \begin{Bmatrix} x \\ y \end{Bmatrix}. \tag{3.21}$$

By derivation we get:

$$\begin{Bmatrix} \dot{x}_1 \\ \dot{y}_1 \end{Bmatrix} = \begin{Bmatrix} \dot{x}_0 \\ \dot{y}_0 \end{Bmatrix} + \dot{\theta} \begin{Bmatrix} -y \\ x \end{Bmatrix}, \tag{3.22}$$

or: $\vec{v}_M = \vec{v}_0 + \vec{\omega} \times \vec{r}$ with the components:

$$v_x = v_{0x} - \omega y; \quad v_y = v_{0y} + \omega x; \quad v_z = 0, \tag{3.23}$$

where $\vec{\omega} = \omega \vec{k}$; $\quad \vec{r} = x\vec{i} + y\vec{j}$.

To determine the acceleration of a point of the rigid body in plane-parallel motion, the relationship obtained in the general motion of the rigid body is used:

$$\vec{a}_M = \vec{a}_0 + \vec{\varepsilon} \times \vec{r} + \vec{\omega} \times (\vec{\omega} \times \vec{r}) = \vec{a}_0 + \vec{\varepsilon} \times \vec{r} + (\vec{\omega}\,\vec{r})\vec{\omega} - \omega^2\vec{r}. \tag{3.24}$$

Taking into account the fact that $\vec{\omega} \perp \vec{r}$ it follows that: $\vec{\omega}\,\vec{r} = 0$, $\vec{\varepsilon} = \dot{\vec{\omega}} = \dot{\omega}\,\vec{k}$.

$$\vec{a}_M = \vec{a}_0 + \vec{\varepsilon} \times \vec{r} - \omega^2\vec{r}, \tag{3.25}$$

with the components:

$$a_x = a_{0x} - \varepsilon y - \omega^2 x; \quad a_y = a_{0y} + \varepsilon x - \omega^2 y; \quad a_z = 0. \tag{3.26}$$

3.4.2 Fixed and Mobile Centroid

The problem of the instantaneous axis of rotation and translation for plane-parallel motion is posed. The minimum speed in the points belonging to A.I.R.T. is $v_{min} = (\vec{\omega}\,\vec{v}_0)/\omega = 0$. In this case we are talking about the instantaneous axis of rotation. Therefore, the points of zero speed are searched for. From the written relation for the velocity components, it follows that the coordinate point ξ, η, ζ with zero velocity must verify the relations:

$$\mathbf{v}_{0x} - \omega\eta = 0; \quad \mathbf{v}_{0y} + \omega\xi = 0; \quad \varsigma = \text{arbitrary}, \tag{3.27}$$

where from:

$$\xi = -\frac{\mathbf{v}_{0y}}{\omega}; \quad \eta = \frac{\mathbf{v}_{0x}}{\omega}; \quad \varsigma = \text{arbitrary}. \tag{3.28}$$

It follows that the points of the zero velocity of the rigid body are found on the line that has, compared to the mobile coordinate system, the coordinates determined with the written relations. The line is only the instantaneous axis of rotation with no translation along it. The rigid body behaves in terms of velocity distribution as if it were rotating instantaneously around this axis (exactly as in fixed-axis rotation).

The geometric place of A.I.R. (instantaneous axis of rotation) in relation to the fixed and mobile reference system, are cylinders, called axoids (see Fig. 3.14).

The point of zero velocity in the *Oxy* coordinate plane is called the *instantaneous center of rotation* (C.I.R.). If we make a translation in the instantaneous center of rotation with the relations:

$$x = x' + \xi; \quad y = y' + \eta; \quad z = z'; \tag{3.29}$$

the expressions for the speed components are obtained:

$$v_x = -\omega y'; \quad v_y = \omega x; \quad v_z = 0, \tag{3.30}$$

relations analogous to those in the rotational motion with a fixed axis.

Definition. The geometric locus of the instantaneous center of rotation in relation to the mobile reference system is called the mobile or rolling centroid. The geometric locus of the instantaneous center of rotation in relation to the fixed reference system is called the fixed centroid or base centroid.

The centroids therefore represent the intersection of the axoids (cylinders) with the coordinate plane. The parametric equations of the mobile centroid are given by:

$$\xi = -\frac{v_{0y}}{\omega}; \quad \eta = \frac{v_{0x}}{\omega}. \tag{3.31}$$

To obtain the parametric equations of the fixed centroid, it is necessary to transform the expressions obtained in the fixed reference system with the relations:

$$\xi_1 = x_0 + \xi \cos\theta - \eta \sin\theta; \quad \eta_1 = y_0 + \xi \sin\theta + \eta \cos\theta. \tag{3.32}$$

The functions ξ, η, ξ_1, η_1 depend on the time parameter or another conveniently chosen time-dependent parameter.

Geometric methods can also be used to determine the instantaneous center of rotation. They have the advantage of simplicity in certain cases. The properties of the speed distribution are used for this purpose. The movement is expressed in terms of speeds like a rotational movement around the I.C.R. The speed of a point is perpendicular to the line that joins that point with the I.C.R. (perpendicular to the radius) and proportional to it. So, if the directions of the velocities of two points A and B in planar motion are known, then I.C.R, is at the intersection of the perpendiculars to the directions of the velocities. The magnitude of the speed of points A and B is given by the relations: $v_A = \omega \cdot IA$ and $v_B = \omega \cdot IB$. It follows that if v_A is known, ω can be found with the relation $\omega = v_A/IA$ (IA measured on the figure) and v_B with the relation $v_B = v_A IB/IA$ (Fig. 3.15). If the supports of the velocities of the two points are parallel, then the two normals have no point of intersection. In this case I.C.R. is at infinity, and the movement is that of planar translation.

A special case is when the directions of the velocities of two points are known, perpendicular to the line joining the two points (Fig. 3.16). In this case, it is necessary to know the values of the speeds. Using the property of linear variation of speeds, the I.C.R. is determined, at the intersection of the line joining the two points with the line joining the extremities of the speeds. Some application in real world of kinematics can be found in [7–17].

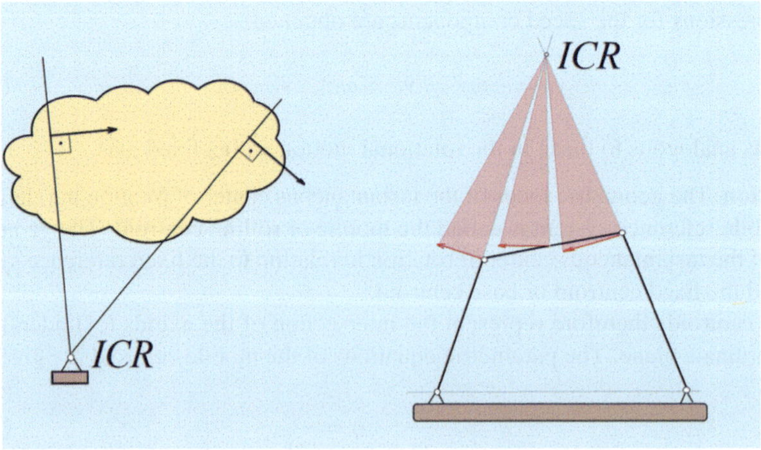

Fig. 3.15 Instantaneous center of rotation (ICR) of a rigid body

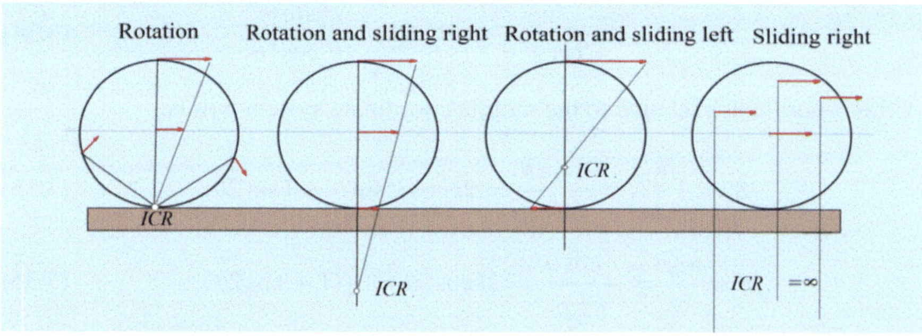

Fig. 3.16 Instantaneous center of rotation (ICR)

Example 1. **Cardan's problem**. A bar of length $2l$ moves so that end A slides on the O_1x_1 axis with constant speed v, and B slide on the O_1y_1 axis. Determine the angular velocity and acceleration of the bar and its centroid (fixed and mobile) (Fig. 3.17).

(i) *Analytical solution.* The components of the velocity on the axes of the mobile reference system are: $v_{0x} = v\cos\theta$; $v_{0y} = -v\sin\theta$ and $\omega = \dot{\theta}$. The speed of point B is given by the relation:

$$\vec{v}_B = \vec{v}_A + \vec{\omega} \times \overrightarrow{AB} = \left(v - 2l\dot{\theta}\cos\theta\right)\vec{i} - 2l\dot{\theta}\sin\theta\,\vec{j}. \tag{3.33}$$

The condition that the speed of point B is oriented along the Oy axis gives: $\dot{\theta} = \omega = \frac{v}{2l\cos\theta}$. The angular acceleration of the bar is obtained by derivation:

Fig. 3.17 A sliding bar

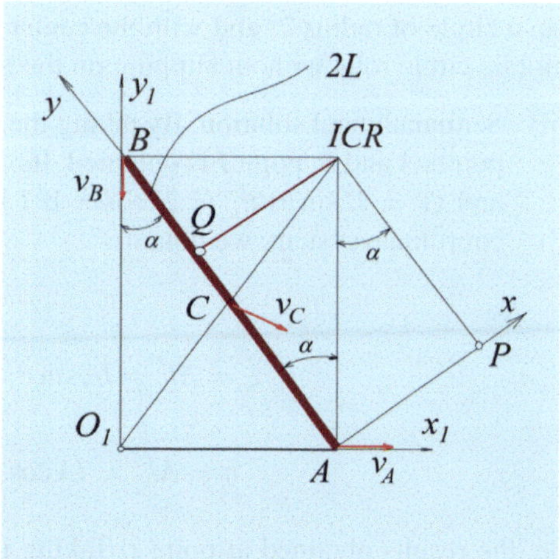

$$\varepsilon = \dot{\omega} = \frac{v}{2l\cos^2\theta}\sin\theta \;\; \dot{\theta} = \frac{v^2\sin\theta}{4l^2\cos^3\theta}. \tag{3.34}$$

C.I.R. coordinates relative to the mobile coordinate system will be:

$$\xi = -\frac{v_{Oy}}{\omega} = \frac{v\sin\theta}{\frac{v}{2l\cos\theta}} = 2l\cos\theta\sin\theta = l\sin 2\theta; \tag{3.35}$$

$$\eta = \frac{v_{Ox}}{\omega} = \frac{v\cos\theta}{\frac{v}{2l\cos\theta}} = 2l\cos^2\theta = \frac{l}{2}(1+\cos 2\theta). \tag{3.36}$$

Eliminating the parameter θ between the two relations taking into account the expression: $\cos^2 2\theta + \sin^2 2\theta = 1$ the mobile centroid equation is obtained:

$$\xi^2 + (\eta - l)^2 = l^2; \tag{3.37}$$

that is, the equation of a circle with its center at the center of the bar and with a radius equal to half the length of the bar.

ICR coordinates relative to the fixed reference system will be given by:

$$\xi_1 = 2l\sin\theta + \xi\cos\theta - \eta\sin\theta = 2l\sin\theta; \tag{3.38}$$

$$\eta_1 = \xi\sin\theta + \eta\cos\theta = 2l\cos\theta. \tag{3.39}$$

Eliminate θ by squaring and adding and obtain the base equation in the form:

$$\xi_1^2 + \eta_1^2 = 4l^2, \tag{3.40}$$

i.e. a circle of radius $2l$ and with the center at the origin O_1. During movement, the mobile circle rolls without slipping on the fixed circle (Fig. 3.18).

(ii) Semianalytical solution. By taking the perpendiculars to the velocity vectors at points A and B, point I is obtained. Its coordinates in the fixed reference system are: $\xi_1 = 2l\sin\theta$; $\eta_1 = 2l\cos\theta$. If I is projected onto the axes of the mobile coordinate system, we obtain:

$$\xi = AI' = IA\sin\theta = 2l\sin\theta\cos\theta; \tag{3.41}$$

$$\eta = AI'' = IA\cos\theta = 21\cos^2\theta, \tag{3.42}$$

i.e. the results obtained at point i. Taking into account that the angular velocity is $v_A = v = \omega\; IA = \omega\; 2l\;\cos\theta$, ω is obtained.

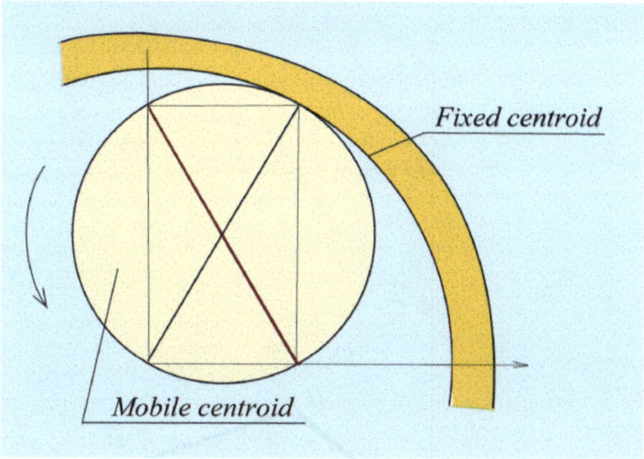

Fig. 3.18 Centroids

(iii) *Geometric solution.* In the case of the construction from point ii), the figure O1AIB is a rectangle. Compared to the bar, angle AIB is always right. Point I will then be on the circle circumscribed by the rectangle of diameter $AB = 2\,l$. Distance $O_1 I = AB = ct$ as diagonals of the rectangle and then relative to the fixed reference system point I moves on the circle of radius equal to $O_1 I = 2\,l$ and with the center in O_1.

Example 2. **Slider-crank mechanism**. We propose to determine the centroidds for the connecting rod in the connecting slider-crank mechanism (Figs. 3.19 and 3.20).

$$x_1 = AC = x_C = r\cos\alpha + l\cos\beta = r\cos\alpha + l\sqrt{1 - (r/l)^2\sin\alpha} \tag{3.43}$$

$$y_1 = IC = x_C tg\alpha; \tag{3.44}$$

$$AI = x_C/\cos\alpha; \quad BI = AI - r; \tag{3.45}$$

$$x = BI\cos(\alpha + \beta); \quad y = BI\sin(\alpha + \beta). \tag{3.46}$$

3.4.3 Accelerations Pole

There is the problem of zero acceleration points. If ξ', η' and ζ are the coordinates of such a point, they will have to meet the conditions:

$$a_{0x} - \varepsilon\eta' - \omega^2\xi' = 0; \quad a_{0y} - \varepsilon\xi' - \omega^2\eta' = 0; \quad \zeta = \text{ arbitrary}. \tag{3.47}$$

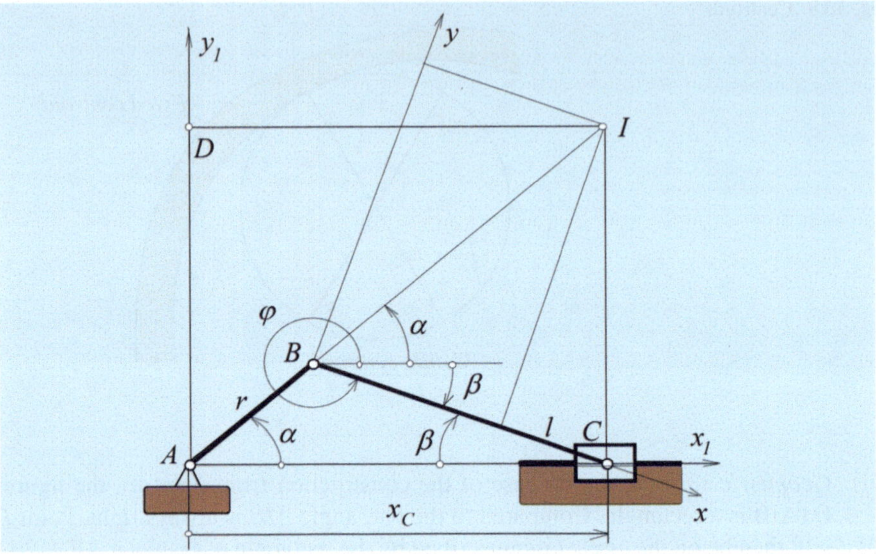

Fig. 3.19 Slider-crank mechanism

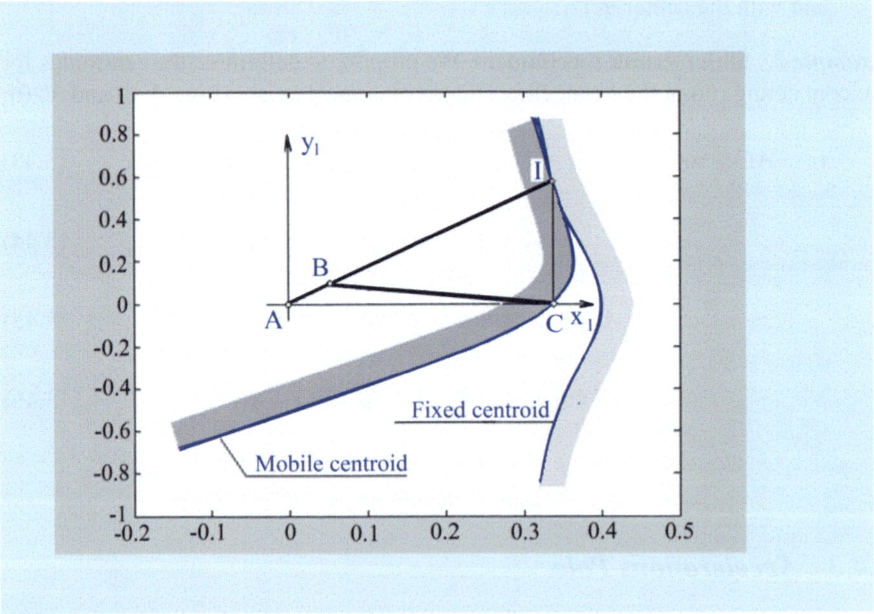

Fig. 3.20 Fixed and mobile centroid. The scales along the two axes are different

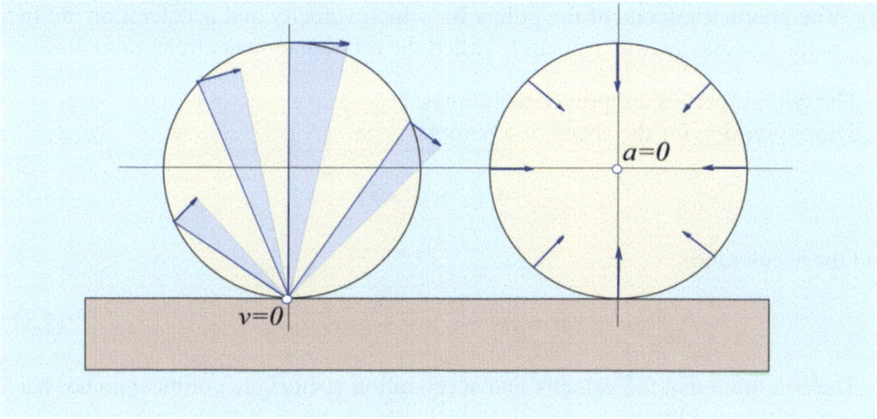

Fig. 3.21 Instantaneous center of rotation (left) and pole of accelerations (right)

It results:

$$\xi' = \frac{\omega^2 a_{0x} - \varepsilon a_{0y}}{\omega^4 + \varepsilon^2}; \quad \eta' = \frac{\omega^2 a_{0y} + \varepsilon a_{0x}}{\omega^4 + \varepsilon^2}; \quad \zeta = \text{arbitrary}. \tag{3.48}$$

The set of zero acceleration points is found on a line perpendicular to the direction plane. The point in the plane of the mobile reference system of zero acceleration and coordinates ξ', η' are called the *acceleration pole* (acceleration center). The coordinates of the acceleration pole with respect to the fixed reference system are obtained as in the case of C.I.R. where ξ, η, ξ_1, η_1 are replaced by ξ', η', ξ_1', η_1'. The distribution of accelerations will be identical to that of the rotation movement with a fixed axis, as if the body were to rotate around the line perpendicular to the plane that passes through the pole of accelerations. As in the case of velocities, if we do the translation, the components of the acceleration are obtained in the form:

$$a_x = -\varepsilon y' - \omega^2 x'; \quad a_y = \varepsilon x' - \omega^2 y'; \quad a_z = 0, \tag{3.49}$$

which are identical to those in the rotation motion.

Figure 3.21 shows the instantaneous center of rotation and the pole of accelerations for a wheel that rolls without slipping on the horizontal plane. Note that the two points do not coincide, in general.

3.4.4 Properties in Plane-Parallel Motion (Bresse Circles)

(a) The geometric locus of the points for which velocity and acceleration are two collinear vectors is a circle called the circle of inflections.

(b) The geometric locus of the points for which velocity and acceleration are two perpendicular vectors is a circle called the reimbursement circle.

The two properties are proved as follows:
The expression for the speed of a point is:

$$\vec{v} = \vec{v}_0 + \vec{\omega} \times \vec{r}, \tag{3.50}$$

and the acceleration:

$$\vec{a} = \vec{a}_0 + \vec{\varepsilon} \times \vec{r} - \omega^2 \vec{r}. \tag{3.51}$$

The condition that the velocity and acceleration vectors are collinear comes back to putting the condition: $\vec{a} \times \vec{v} = 0$, and the condition that the two vectors are perpendicular comes back to: $\vec{a} \cdot \vec{v} = 0$. Performing the calculations, in both cases, the equation of a circle is obtained.

3.5 Rigid Motion with Fixed Point

3.5.1 Expressions of Velocity and Acceleration

It is considered a rigid body which during the movement has fixed a point O (Fig. 3.22). The origin of the mobile and fixed reference systems is chosen at point O. The velocity and acceleration of this point are zero. The velocity of a certain point of the rigid body is given by the relation:

$$\vec{v}_M = \vec{v}_0 + \vec{\omega} \times \vec{r} = \vec{\omega} \times \vec{r}. \tag{3.52}$$

The components of velocity are:

$$\begin{aligned} v_x &= \omega_y z - \omega_z y; \\ v_y &= \omega_z x - \omega_x z; \\ v_z &= \omega_x y - \omega_y x. \end{aligned} \tag{3.53}$$

Zero velocity points are searched for. The vector expression for velocity indicates that points on the line $\vec{r} = \lambda \vec{\omega}$ have zero velocity. This line will be the instantaneous axis of rotation. The velocity distribution will be identical to that in fixed-axis rotational motion, as if the rigid body were rotating about the instantaneous axis of rotation.

The instantaneous axis of rotation and translation can also be determined with the relation obtained in Chap. 2:

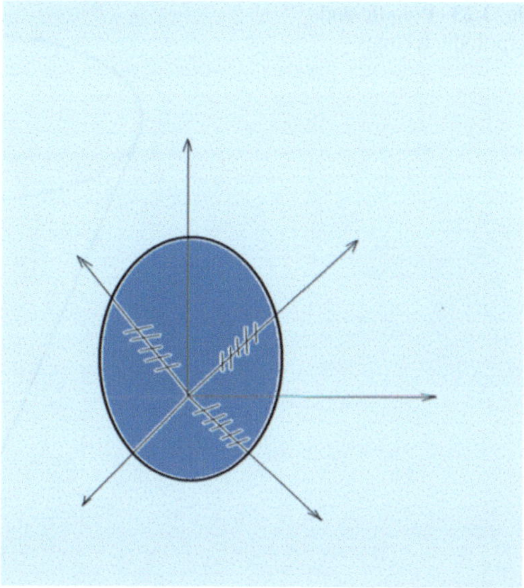

Fig. 3.22 Rigid body with fixed point

$$\vec{r} = \frac{\vec{\omega}\vec{v}_0}{\omega^2} + \lambda\vec{\omega} = \lambda\vec{\omega}, \tag{3.54}$$

so a straight line passing through O. All points on this axis have the speed:

$$v_{\min} = \frac{\vec{v}_0\vec{\omega}}{\omega} = 0, \tag{3.55}$$

so it is an axis of rotation without translation. Because at any moment, the A.I.R. passes through point O, it follows:

- the geometric locus of A.I.R. relative to the fixed reference system is a cone with the apex in O_1 called the *herpolodic cone*;
- the geometric locus of A.I.R. in relation to the mobile reference system, it is also a cone with apex in O called the *polodic cone* (Fig. 3.23).

The two cones are called *Poinsot's cones.* During the motion, the polodic cone rolls without slipping on the fixed herpolodic cone. The two cones are tangent.

The acceleration is given by the relation:

$$\vec{a}_M = \vec{a}_0 + \vec{\varepsilon} \times \vec{r} + \vec{\omega} \times (\vec{\omega} \times \vec{r}) = \vec{\varepsilon} \times \vec{r} + \vec{\omega} \times (\vec{\omega} \times \vec{r}), \tag{3.56}$$

with the components:

$$a_x = -\left(\omega_y^2 + \omega_z^2\right)x + \left(\omega_x\omega_y - \varepsilon_z\right)y + \left(\omega_x\omega_z + \varepsilon_y\right)z\ ; \tag{3.57}$$

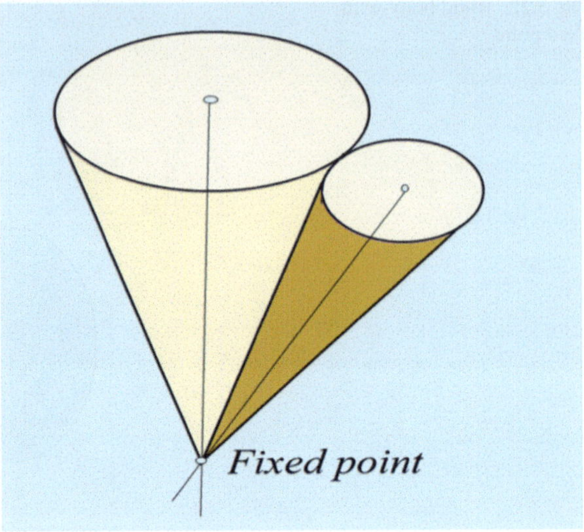

Fig. 3.23 Polodic and herpolodic a cones

$$a_y = -\left(\omega_z^2 + \omega_x^2\right)y + \left(\omega_y\omega_z - \varepsilon_x\right)z + \left(\omega_y\omega_x + \varepsilon_z\right)x; \tag{3.58}$$

$$a_z = -\left(\omega_x^2 + \omega_y^2\right)z + \left(\omega_z\omega_x - \varepsilon_y\right)x + \left(\omega_z\omega_y + \varepsilon_x\right)y. \tag{3.59}$$

In order to determine the distribution of accelerations we make the observation that if in the relation expressing the acceleration of a point we change $\vec{r}$ on with $\lambda\vec{r}$ it results:

$$\vec{a} = \vec{\varepsilon} \times (\lambda\vec{r}) + \vec{\omega} \times (\vec{\omega} \times \lambda\vec{r}) = \lambda[\vec{\varepsilon} \times \vec{r} + \vec{\omega} \times (\vec{\omega} \times \vec{r})] = \lambda\vec{a}.$$

It follows that along a straight line passing through the origin, the acceleration vector varies linearly. We look for other zero acceleration points outside the fixed point, thus putting the condition $\vec{a} = 0$ we get the system:

$$0 = -\left(\omega_y^2 + \omega_z^2\right)x + \left(\omega_x\omega_y - \varepsilon_z\right)y + \left(\omega_x\omega_z + \varepsilon_y\right)z\ ; \tag{3.60}$$

$$0 = -\left(\omega_z^2 + \omega_x^2\right)y + \left(\omega_y\omega_z - \varepsilon_x\right)z + \left(\omega_y\omega_x + \varepsilon_z\right)x; \tag{3.61}$$

$$0 = -\left(\omega_x^2 + \omega_y^2\right)z + \left(\omega_z\omega_x - \varepsilon_y\right)x + \left(\omega_z\omega_y + \varepsilon_x\right)y; \tag{3.62}$$

which is linearly homogeneous with the determinant:

$$\begin{bmatrix} -\left(\omega_y^2+\omega_z^2\right) & \omega_x\omega_y-\varepsilon_z & \omega_x\omega_z+\varepsilon_y \\ \omega_y\omega_x+\varepsilon_z & -\left(\omega_z^2+\omega_x^2\right) & \omega_y\omega_z-\varepsilon_x \\ \omega_z\omega_x-\varepsilon_y & \omega_z\omega_y+\varepsilon_x & -\left(\omega_x^2+\omega_y^2\right) \end{bmatrix}$$
$$=-\left[\left(\omega_y\varepsilon_z-\omega_z\varepsilon_y\right)^2+\left(\omega_z\varepsilon_x-\omega_x\varepsilon_z\right)^2+\left(\omega_x\varepsilon_y-\omega_y\varepsilon_x\right)^2\right]=-(\vec{\omega}\times\vec{\varepsilon})^2.$$

$$\begin{bmatrix} -\left(\omega_y^2+\omega_z^2\right) & \omega_x\omega_y-\varepsilon_z & \omega_x\omega_z+\varepsilon_y \\ \omega_y\omega_x+\varepsilon_z & -\left(\omega_z^2+\omega_x^2\right) & \omega_y\omega_z-\varepsilon_x \\ \omega_z\omega_x-\varepsilon_y & \omega_z\omega_y+\varepsilon_x & -\left(\omega_x^2+\omega_y^2\right) \end{bmatrix}$$
$$=-\left[\left(\omega_y\varepsilon_z-\omega_z\varepsilon_y\right)^2+\left(\omega_z\varepsilon_x-\omega_x\varepsilon_z\right)^2+\left(\omega_x\varepsilon_y-\omega_y\varepsilon_x\right)^2\right]$$
$$=-(\vec{\omega}\times\vec{\varepsilon})^2. \tag{3.63}$$

In general $\Delta \neq 0$, so the point of zero acceleration is the origin O, $x = y = z = 0$. If $\vec{\omega}$ and $\vec{\varepsilon}$ are collinear, then $\Delta = 0$, so there is a line on which the acceleration is zero (for example, the case fixed axis rotation).

3.5.2 Euler's Angles

The rotation of the mobile reference system relative to the fixed reference system is characterized by the nine principal cosines of the rotation matrix $[R]$ which have been seen to be linked by six relations. Three independent cosines can therefore be chosen to describe the rotation of the rigid body, the other six can be expressed as a function of them. Such a way of studying the problem is particularly cumbersome. There can be various ways to choose three independent parameters to describe the finite rotation of the rigid body, the nine cosine directors being expressed as a function of them.

A much more used way, especially due to its intuitive support, is to use Euler's angles as independent parameters. In this way, the global rotation is obtained by means of three planar rotations, successively applied to the mobile reference system. Representing the fixed and the mobile reference system, Euler's angles are obtained in the following way: intersect the Oxy plane with the $O_1x_1y_1$ plane. The intersection line is ON. The angle ψ made by Ox_1 with ON is called the precession angle, and the angle φ made by Ox with ON is called the intrinsic rotation angle. The third parameter that will define the position of the referential $Oxyz$ is the angle θ, called the nutation angle. It is presented how the three previously defined angles can be used as independent parameters for defining the rotation matrix.

- we consider the fixed reference system $O_1x_1y_1z_1$ and the reference system that rotated together with the rigid one, $Oxyz$ (Fig. 3.24). The transition from the fixed reference system to the rotated one is done in the following three steps:

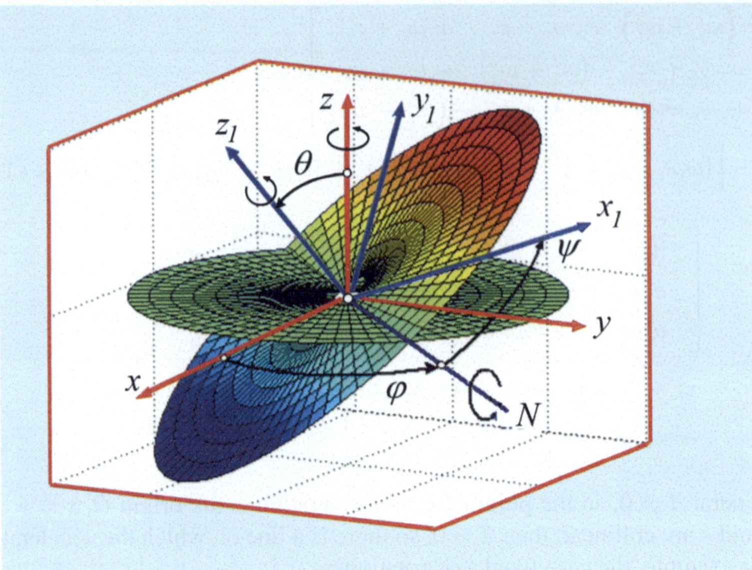

Fig. 3.24 The Euler's angles

- a planar rotation of angle ψ of the reference system is performed around the axis Oz_1 considered fixed; the reference system has a new position defined by the axes $Ox'y'z'$, $Oz \equiv Oz'$. The Ox' axis is also denoted ON and is called the node axis;
- the axis $ON \equiv Ox'$ is kept fixed and a planar rotation of angle θ is performed around it; the reference system has a new position defined by the axes $Ox''y''z''$, $Ox' \equiv Ox''$;
- the Oz'' axis is kept fixed and the reference system is rotated around it by the angle φ; the reference system that defines the final position of the rigid $Oxyz$, with $Oz'' \equiv Oz$ is obtained.

For the first operation we therefore have a plane rotation of angle ψ which will be defined by the rotation matrix $[\boldsymbol{R}_\psi]$. The link between the coordinates of a vector expressed in the two reference systems $O_1x_1y_1z_1$ and $Ox'y'z'$ is:

$$\begin{Bmatrix} x_1 \\ y_1 \\ z_1 \end{Bmatrix} = \begin{bmatrix} \cos\psi & -\sin\psi & 0 \\ \sin\psi & \cos\psi & 0 \\ 0 & 0 & 1 \end{bmatrix} \begin{Bmatrix} x' \\ y' \\ z' \end{Bmatrix}, \tag{3.64}$$

or:

$$\{r_1\} = [R_\psi]\{r'\}$$

The second rotation around the axis of the nodes, of angle θ is expressed by the rotation matrix $[\boldsymbol{R}_\theta]$, and the connection between the components of a vector expressed in the two coordinate systems is given by the relation:

$$\begin{Bmatrix} x' \\ y' \\ z' \end{Bmatrix} = \begin{bmatrix} 1 & 0 & 0 \\ 0 & \cos\theta & -\sin\theta \\ 0 & \sin\theta & \cos\theta \end{bmatrix} \begin{Bmatrix} x'' \\ y'' \\ z'' \end{Bmatrix} \tag{3.65}$$

or:

$$\{r'\} = [R_\theta]\{r''\}.$$

The third rotation around the axis $Oz = Oz''$, of angle φ is expressed by the rotation matrix $[\boldsymbol{R}_\varphi]$, and the connection between the components of a vector expressed in the two coordinate systems is given by the relation:

$$\begin{Bmatrix} x'' \\ y'' \\ z'' \end{Bmatrix} = \begin{bmatrix} \cos\varphi & -\sin\varphi & 0 \\ \sin\varphi & \cos\varphi & 0 \\ 0 & 0 & 1 \end{bmatrix} \begin{Bmatrix} x \\ y \\ z \end{Bmatrix} \tag{3.66}$$

or

$$\{r''\} = \lfloor R_\phi \rfloor \{r\}.$$

Finally one can write:

$$\{r_1\} = [R_\psi]\{r'\} = [R_\psi][R_\theta]\{r''\} = [R_\psi][R_\theta][R_\phi]\{r\}. \tag{3.67}$$

The matrix representing the rotation is:

$$[R] = [R_\psi][R_\theta][R_\phi]. \tag{3.68}$$

Doing the calculations we get:

$$[R] = \begin{bmatrix} \cos\psi\cos\phi - \sin\psi\sin\phi\cos\theta & -\cos\psi\sin\phi - \sin\psi\cos\phi\cos\theta & \sin\psi\sin\theta \\ \sin\psi\cos\phi + \cos\psi\sin\phi\cos\theta & -\sin\psi\sin\phi + \cos\psi\cos\phi\cos\theta & -\sin\theta\cos\phi \\ \sin\theta\sin\psi & -\sin\theta\cos\psi & \cos\theta \end{bmatrix}$$
$$= \begin{bmatrix} \{i\} & \{j\} & \{k\} \end{bmatrix}. \tag{3.69}$$

The lines $\{i\}$, $\{j\}$, $\{k\}$ have the components, expressed in the fixed coordinate system:

$$\{i\} = \begin{Bmatrix} \cos\psi\cos\phi - \sin\psi\sin\phi\cos\theta \\ \sin\psi\cos\phi + \cos\psi\sin\phi\cos\theta \\ \sin\theta\sin\psi \end{Bmatrix}; \tag{3.70}$$

$$\{j\} = \begin{Bmatrix} -\cos\psi\sin\phi - \sin\psi\cos\phi\cos\theta \\ -\sin\psi\sin\phi + \cos\psi\cos\phi\cos\theta \\ -\sin\theta\cos\psi \end{Bmatrix}; \tag{3.71}$$

$$\{k\} = \begin{Bmatrix} \sin\psi\sin\theta \\ -\sin\theta\cos\phi \\ \cos\theta \end{Bmatrix}. \tag{3.72}$$

If we consider that the final rotation is obtained by the sequence of three rotations of angles ψ, θ, φ in a very small time interval Δt and if we consider that the ratios between angles and time represent the angular speeds of rotation around the three axes, then the angular velocity of the rigid body can be represented as the vector sum of the three angular velocities:

$$\{\omega\} = \dot{\psi}\{k_1\} + \dot{\theta}\{e'\} + \dot{\phi}\{k\}. \tag{3.73}$$

Taking into account that: $\{e'\} = \{i_1\}\cos\psi + \{j_1\}\sin\psi$ and {k} is previously defined within the rotation matrix we have the angular velocity components in the fixed reference system:

$$\omega_{x1} = \dot{\theta}\cos\psi + \dot{\phi}\sin\theta\sin\psi\ ; \tag{3.74}$$

$$\omega_{y1} = \dot{\theta}\sin\psi - \dot{\phi}\sin\theta\cos\psi\ ; \tag{3.75}$$

$$\omega_{z1} = \dot{\psi} + \dot{\phi}\cos\theta\ . \tag{3.76}$$

Of course, these components could be obtained if the relation defined in Chap. 2 was applied for the angular velocity operator. In this case, however, a large number of calculations would have to be performed. The components of the angular velocity in the local reference system are obtained with the relations:

$$\{\Omega\} = [R]^T\{\omega\}, \tag{3.77}$$

where from:

$$\omega_x = \dot{\psi}\sin\theta\sin\phi + \dot{\theta}\cos\phi\ ; \tag{3.78}$$

$$\omega_y = \dot{\psi}\sin\theta\cos\phi - \dot{\theta}\sin\phi\ ; \tag{3.79}$$

$$\omega_z = \dot{\phi} + \dot{\psi}\cos\theta\ . \tag{3.80}$$

3.5.3 Bryan-Cardan Angles

The transition from one reference system to the other is done through three plane rotations in the following way: the first angular rotation Φ_1 is done around the Ox_1 axis. The new reference system will be denoted $Ox'y'z'$. Now a second rotation is made, of an angle Φ_2 around the axis Oy'. The reference system becomes $Ox''y''z''$. Finally, an angular rotation Φ_3 is made around the Ox'' axis. The final position of the $Oxyz$ reference system is thus obtained. Let us denote the plane rotation matrices in the three cases by $[R_1]$, $[R_2]$, $[R_3]$. Since we are dealing with plane rotations we have:

$$[R_1] = \begin{bmatrix} 1 & 0 & 0 \\ 0 & \cos\Phi_1 & -\sin\Phi_1 \\ 0 & \sin\Phi_1 & \cos\Phi_1 \end{bmatrix}; \tag{3.81}$$

$$[R_2] = \begin{bmatrix} \cos\Phi_2 & 0 & \sin\Phi_1 \\ 0 & 1 & 0 \\ -\sin\Phi_1 & 0 & \cos\Phi_2 \end{bmatrix}; \tag{3.82}$$

$$[R_3] = \begin{bmatrix} \cos\Phi_3 & -\sin\Phi_3 & 0 \\ \sin\Phi_1 & \cos\Phi_3 & 0 \\ 0 & 0 & 1 \end{bmatrix}. \tag{3.83}$$

The transformation of the coordinates of a vector in the different coordinate systems is done according to the relations:

$$\{r\}_{Ox1y1x1} = [R_1]\{r\}_{Ox'y'z'}; \tag{3.84}$$

$$\{r\}_{Ox'y'z'} = [R_2]\{r\}_{Ox''y''z''}; \tag{3.85}$$

$$\{r\}_{Ox''y''z''} = [R_3]\{r\}_{Oxyz}. \tag{3.86}$$

It results:

$$\begin{aligned} \{r\}_{Ox1y1x1} &= [R_1][R_2][R_3]\{r\}_{Oxyz} \\ &= \begin{bmatrix} 1 & 0 & 0 \\ 0 & \cos\Phi_1 & -\sin\Phi_1 \\ 0 & \sin\Phi_1 & \cos\Phi_1 \end{bmatrix} \begin{bmatrix} \cos\Phi_2 & 0 & \sin\Phi_2 \\ 0 & 1 & 0 \\ -\sin\Phi_2 & 0 & \cos\Phi_2 \end{bmatrix} \begin{bmatrix} \cos\Phi_3 & -\sin\Phi_3 & 0 \\ \sin\Phi_3 & \cos\Phi_3 & 0 \\ 0 & 0 & 1 \end{bmatrix} \\ &= \begin{bmatrix} \cos\Phi_2 & 0 & \sin\Phi_2 \\ \sin\Phi_1\sin\Phi_2 & \cos\Phi_1 & -\sin\Phi_1\cos\Phi_2 \\ -\sin\Phi_2\cos\Phi_1 & \sin\Phi_1 & \cos\Phi_1\cos\Phi_2 \end{bmatrix} \begin{bmatrix} \cos\Phi_3 & -\sin\Phi_3 & 0 \\ \sin\Phi_3 & \cos\Phi_3 & 0 \\ 0 & 0 & 1 \end{bmatrix} \end{aligned}$$

$$= \begin{bmatrix} \cos\Phi_2\cos\Phi_3 & -\cos\Phi_2\sin\Phi_3 & \sin\Phi_2 \\ \sin\Phi_1\sin\Phi_2\cos\Phi_3+\cos\Phi_1\sin\Phi_3 & -\sin\Phi_1\sin\Phi_2\sin\Phi_3+\cos\Phi_1\cos\Phi_3 & -\sin\Phi_1\cos\Phi_2 \\ -\sin\Phi_1\cos\Phi_1\cos\Phi_3+\sin\Phi_1\sin\Phi_3 & \sin\Phi_2\sin\Phi_3\cos\Phi_1+\sin\Phi_1\cos\Phi_3 & \cos\Phi_1\cos\Phi_2 \end{bmatrix} \tag{3.87}$$

where the final matrix obtained is the rotation matrix, from the mobile reference system Oxyz to the fixed reference system $Ox_1y_1z_1$. For small angles $\Phi_1 \leq 1$, $\Phi_2 \leq 1$, $\Phi_3 \leq 1$ approximations $\sin\Phi_i \approx \Phi_i$, $\cos\Phi_i \approx 1$ can be made, and the rotation matrix becomes:

$$[R] = \begin{bmatrix} 1 & -\Phi_3 & \Phi_2 \\ \Phi_3 & 1 & -\Phi_1 \\ -\Phi_2 & \Phi_1 & 1 \end{bmatrix} = [E] - [\Phi], \tag{3.88}$$

where $[E]$ is the unit matrix and $[\Phi]$ the skewsymmetric matrix:

$$[\Phi] = \begin{bmatrix} 0 & -\Phi_3 & \Phi_2 \\ \Phi_3 & 0 & -\Phi_1 \\ -\Phi_2 & \Phi_1 & 0 \end{bmatrix}. \tag{3.89}$$

The skewsymmetric matrix $[\Phi]$ can represent a vector product. If we consider that the rotation takes place in a very small time interval and we divide these angles by this very small time interval, we obtain the angular velocity with three components: $\dot{\Phi}_1, \dot{\Phi}_2, \dot{\Phi}_3$ the angular velocity operator:

$$[\omega] = \begin{bmatrix} 0 & -\dot{\Phi}_3 & \dot{\Phi}_2 \\ \dot{\Phi}_3 & 0 & -\dot{\Phi}_1 \\ -\dot{\Phi}_2 & \dot{\Phi}_1 & 0 \end{bmatrix}; \tag{3.90}$$

and the angular velocity vector:

$$\{\omega\} = \begin{Bmatrix} \dot{\Phi}_1 \\ \dot{\Phi}_2 \\ \dot{\Phi}_3 \end{Bmatrix}. \tag{3.91}$$

This vector is an eigenvector for the rotation matrix since:

$$[R]\{\omega\} = ([E] - [\Phi])\{\Phi\} = \{\Phi\}. \tag{3.92}$$

This means that the vector $\{\Phi\}$, so $\left\{\dot{\Phi}\right\} = \{\omega\}$ have the same components both in the fixed reference system and in the mobile reference system. If the $Oxyz$ reference system makes a rotation defined by the matrix $[\Phi]$ one arbitrary vector transforms according to the relation:

$$\{r\}_{Ox1y1z1} = \{r\}_{Oxyz} + [\Phi]\{r\}_{Oxyz}. \tag{3.93}$$

3.5.4 Rotation Matrix

We have seen that the determinant of the rotation matrix has the value 1 and the inverse of this matrix is equal to its transpose. Let us analyze the problem of vectors and eigenvalues for this matrix $[R]\{v\} = \lambda\{v\}$ or:

$$([R] - \lambda[E])\{v\} = 0. \tag{3.94}$$

The rotation matrix is an orthogonal matrix, so it does not change the lengths. This means that if $\{v\}$ it is an eigenvector, we must therefore have $[R]\{v\} = \{v\}$ so the absolute value of the three eigenvalues must be 1. If the characteristic equation written above is developed, the equation is obtained:

$$\lambda^3 - I_1\lambda^2 + I_2\lambda - I_3 = 0; \tag{3.95}$$

where:

$$I_1 = \text{tr}[R] = e_{11} + e_{22} + e_{33}; \tag{3.96}$$

$$I_2 = (e_{11}e_{22} - e_{12}e_{21}) + (e_{22}e_{33} - e_{23}e_{32}) + (e_{33}e_{11} - e_{31}e_{13}); \tag{3.97}$$

$$I_3 = \det[R] = 1. \tag{3.98}$$

Putting the condition that $\lambda = 1$ is a solution of the equation, the relation is obtained:

$$1 - I_1 + I_2 - 1 = 0; \tag{3.99}$$

or:

$$I_1 = I_2 = \text{tr}[R]; \tag{3.100}$$

so the characteristic equation takes the form:

$$\lambda^3 - \lambda^2\text{tr}[R] + \lambda\text{tr}[R] - 1 = 0. \tag{3.101}$$

If we eliminate the solution $\lambda = 1$, we get:

$$\lambda^2 - \lambda(\text{tr}[R] - 1) + 1 = 0; \tag{3.102}$$

which give us the other two solutions for the eigenvalues:

$$\lambda_{2,3} = \frac{\mathrm{tr}[R] - 1}{2} \pm i\sqrt{1 - \left(\frac{\mathrm{tr}[R] - 1}{2}\right)^2} = \cos\phi \pm i \sin\phi = e^{\pm i\phi}. \qquad (3.103)$$

The notation could be made:

$$\cos\phi = \frac{\mathrm{tr}[R] - 1}{2}; \quad \text{so:} \quad \sin\phi = \sqrt{1 - \left(\frac{\mathrm{tr}[R] - 1}{2}\right)^2} \qquad (3.104)$$

since the elements of the matrix $[R]$, being cosines directors, are subunits, so: $|a_{ii}| \leq 1$. Then $|a_{11} + a_{22} + a_{33}| \leq |a_{11}| + |a_{22}| + |a_{33}| \leq 3$ and $\frac{\mathrm{tr}[R]-1}{2} \leq 1$.

$$|a_{11} + a_{22} + a_{33}| \leq |a_{11}| + |a_{22}| + |a_{33}| \leq 3 \text{ and } \frac{\mathrm{tr}[R] - 1}{2} \leq 1. \qquad (3.105)$$

The particular cases where the rotation matrix is the unit matrix leads to the triple eigenvalue $\lambda = 1$ and the case where the determinant of the rotation matrix is -1 leads to the double eigenvalue $\lambda_{2,3} = -1$.

Let now be $\{n\}$ the normalized eigenvector corresponding to the eigenvalue $\lambda = 1$. It will be the solution of the system:

$$([R] - [E])\{n\} = 0 \,\text{and}\, n_1^2 + n_2^2 + n_3^2 = 1. \qquad (3.106)$$

The vector $\{n\}$ thus obtained will have the same coordinates in the two reference systems. Suppose we make a rotation of the mobile reference system around an axis defined by this vector. For each value of the rotation angle φ corresponds to a position of the fixed reference system relative to the mobile one. It will be shown that any configuration of the mobile reference system can be obtained as a rotation of the system around the eigenvector $\{n\}$ with a certain angle φ. Knowing the rotation matrix will provide the eigenvector and the rotation angle.

We give, without proof, the following theorem:

Theorem (Euler): The displacement of a reference system from an initial position to a final position can be obtained by a rotation of a certain angle, around an axis that remains fixed in both reference systems. The rotation axis is the eigenvector associated with the eigenvalue $\lambda = 1$ of the rotation matrix.

Further, let us determine the rotation matrix, if we know the direction of the axis around which the rotation takes place and we know the angle of rotation φ.

$$\vec{r}^* = \vec{r} + (1 - \cos\phi)\vec{b} + \sin\phi\,\vec{a}; \qquad (3.107)$$

where:

$$\vec{a} = \vec{n}\, x\, \vec{r}; \quad \vec{b} = \vec{n}\, x\, \vec{a}. \qquad (3.108)$$

$$.\vec{r}^* = \vec{r} + (1 - \cos\varphi)\vec{n}x(\vec{n}x\vec{r}) + \sin\varphi\vec{n}x\vec{r}. \tag{3.109}$$

If the matrix notation is adopted, it results:

$$\{r^*\} = \left[\cos\phi\,[E] + (1 - \cos\phi)\{n\}\{n\}^T + \sin\phi[n]\right]\{r\} = [R]\{r\}; \tag{3.110}$$

from where, by identification, the rotation matrix results:

$$\begin{aligned}[R] &= [\cos\varphi[E] + (1 - \cos\varphi)\{n\}\{n\}^T + \sin\varphi[n]] \\ &= \begin{bmatrix} n_1^2 + \left(1 - n_1^2\right)c & n_1n_2(1-c) + n_3s & n_1n_3(1-c) - n_2s \\ n_1n_2(1-c) - n_3s & n_2^2 + \left(1 - n_2^2\right)c & n_2n_3(1-c) + n_1s \\ n_1n_3(1-c) + n_2s & n_2n_3(1-c) - n_1s & n_3^2 + \left(1 - n_3^2\right)c \end{bmatrix}\end{aligned} \tag{3.111}$$

If the rotation angle φ is small ($\phi \ll 1$), the approximations $\sin\phi \approx \phi$; $\cos\phi = 1 - \frac{\phi^2}{2}$ result an approximation for the rotation matrix:

$$[R] = [E] - \phi[n] + \frac{1}{2}\phi^2\left(\{n\}\{n\}^T - [E]\right). \tag{3.112}$$

Linear approximation: $[R] = [E] - \phi[n]$ arose from the study of Bryant-Cardan angles.

3.5.5 *Euler-Rodrigues Parameters*

We have the trigonometric relations:

$$1 - \cos\phi = 2\sin^2\frac{\phi}{2}\ ; \quad \sin\phi = 2\sin\frac{\phi}{2}\cos\frac{\phi}{2}. \tag{3.113}$$

Let us make the notations:

$$q_o = \cos\frac{\phi}{2}\ ; \quad \{q\} = \{n\}\sin\frac{\phi}{2}. \tag{3.114}$$

The vector $\{q\}$ has the direction of the axis of rotation. It will have the same coordinates both in the fixed system and in the mobile system. The coordinates of this vector are denoted by q_1, q_2, q_3.

It turns out that we have the relations:

$$q_o^2 + q_1^2 + q_2^2 + q_3^2 = 1; \tag{3.115}$$

and:

$$1 - 2\{q\}^T\{q\} = q_o^2 - \{q\}^T\{q\} = 2q_o^2 - 1. \tag{3.116}$$

We can also write:

which results in the rotation matrix with elements written according to the Euler-Rodrigues parameters:

$$[R] = \begin{bmatrix} q_o^2 + q_1^2 - q_2^2 - q_3^2 & 2(q_1q_2 + q_oq_3) & 2(q_1q_3 - q_oq_2) \\ 2(q_1q_2 - q_oq_3) & q_o^2 + q_2^2 - q_3^2 - q_1^2 & 2(q_2q_3 + q_oq_1) \\ 2(q_1q_3 + q_oq_2) & 2(q_2q_3 - q_oq_1) & q_o^2 + q_3^2 - q_1^2 - q_2^2 \end{bmatrix}. \tag{3.117}$$

For diagonal elements the alternative form can be given:

$$e_{ii} = 2\left(q_o^2 + q_i^2\right) - 1 \quad (i = 1, 2, 3). \tag{3.118}$$

Depending on the elements of the rotation matrix written in terms of director cosines, the Euler-Rodrigues parameters can be obtained by the relations:

$$q_o = \frac{1}{2}\sqrt{1 + \mathrm{tr}[R]}; \quad q_i = \frac{e_{jk} - e_{kj}}{4q_o}. \tag{3.119}$$

3.5.6 The Connection Between Euler-Rodrigues Parameters and Euler's Angles

We have the relation:

$$\begin{aligned} 1 + \mathrm{tr}[R] &= 1 + c_\theta + c_\theta\left(c_\psi c_\varphi - s_\psi s_\varphi\right) + c_\psi c_\varphi - s_\psi s_\varphi) \\ &= (1 + c_\theta)\left[1 + \left(c_\psi c_\phi - s_\psi s_\phi\right)\right] = (1 + c_\theta)\left(1 + c_{\psi+\phi}\right) \\ &= 4\cos^2\frac{\theta}{2}\cos^2\frac{\psi + \phi}{2}. \end{aligned} \tag{3.120}$$

From here it follows: $q_o = \cos\frac{\theta}{2}\cos\frac{\psi+\varphi}{2}$ the sign can be chosen arbitrarily. For the other parameters, the relations are used:

$$e_{23} - e_{32} = s_\theta\left(c_\psi + c_\varphi\right) = 4\sin\frac{\theta}{2}\cos\frac{\theta}{2}\cos\frac{\psi + \varphi}{2}\cos\frac{\psi - \varphi}{2}; \tag{3.121}$$

where from: $q_1 = \sin\frac{\theta}{2}\cos\frac{\psi-\varphi}{2}$. In the same way one obtains q_2 and q_3.

It results:

$$q_0 = \cos\frac{\theta}{2}\cos\frac{\psi + \phi}{2};$$

$$q_1 = \sin\frac{\theta}{2}\cos\frac{\psi-\phi}{2};\ q_2 = \sin\frac{\theta}{2}\cos\frac{\psi-\phi}{2};\ q_3 = \cos\frac{\theta}{2}\sin\frac{\psi+\phi}{2}.$$

$$q_0 = \cos\frac{\theta}{2}\cos\frac{\psi+\phi}{2};\ q_1 = \sin\frac{\theta}{2}\cos\frac{\psi-\phi}{2};$$
$$q_2 = \sin\frac{\theta}{2}\cos\frac{\psi-\phi}{2};\ q_3 = \cos\frac{\theta}{2}\sin\frac{\psi+\phi}{2}. \tag{3.122}$$

From these formulas also result the relationships:

$$\cos^2\frac{\theta}{2} = q_0^2 + q_3^2,\ \sin^2\frac{\theta}{2} = q_1^2 + q_2^2;\ \mathrm{tg}\frac{\psi+\phi}{2} = \frac{q_3}{q_0};\ \mathrm{tg}\frac{\psi-\phi}{2} = \frac{q_2}{q_1}.$$

$$\cos^2\frac{\theta}{2} = q_0^2 + q_3^2,\ \sin^2\frac{\theta}{2} = q_1^2 + q_2^2;$$
$$tg\frac{\psi+\phi}{2} = \frac{q_3}{q_0};\ tg\frac{\psi-\phi}{2} = \frac{q_2}{q_1}. \tag{3.123}$$

Applications

A1. (a) The locus of points for which velocity and acceleration are two collinear vectors is a circle called the circle of inflections.

(b) The locus of points for which velocity and acceleration are two perpendicular vectors is a circle called the circle of reimbursement (Bresse circles).

Solution: The two properties are demonstrated as follows:

The expression for the velocity of a point is:

$$\vec{v} = \vec{v}_0 + \vec{\omega} \times \vec{r}; \tag{3.124}$$

and of the acceleration:

$$\vec{a} = \vec{a}_0 + \vec{\varepsilon} \times \vec{r} - \omega^2\vec{r}. \tag{3.125}$$

Performing the calculations yields, in both cases, the equation of a circle. The condition that the velocity and acceleration vectors are collinear is equivalent to setting the condition: $\vec{a} \times \vec{v} = 0$, and the condition that the two vectors are perpendicular is equivalent to: $\vec{a} \cdot \vec{v} = 0$.

Performing the calculations, we obtain, in both cases, the equation of a circle.

A2. A bar of length $2L$ (Fig. 3.25) moves in a cylindrical cavity with the speed of the supported end constant $v_A = v_o$. Determine the fixed and mobile centroids of the bar, the angular velocity and angular acceleration. Determine the velocity and acceleration of point B.

Solution: The instantaneous center of rotation I is located at the intersection of the perpendiculars to the direction of the velocities at the corresponding points. The

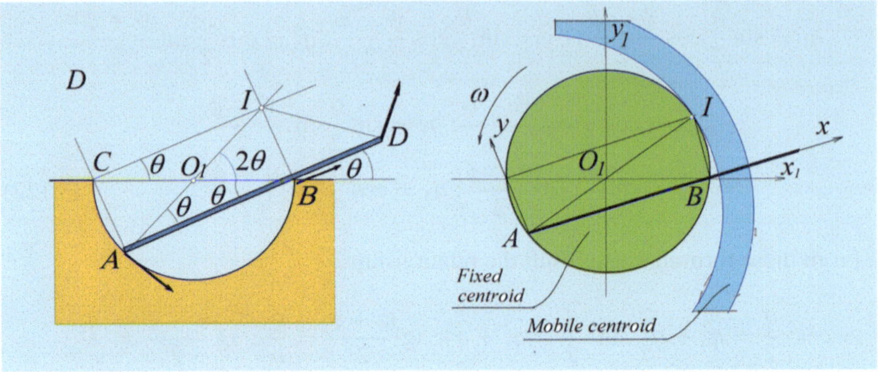

Fig. 3.25 Fixed and mobile centroid

velocity of the point on the bar, which comes into contact with the fixed point B, is along the bar.

The right-angled triangle ABI is inscribable in a circle whose diameter is $O_1I=O_1A=O_1B=R$, with the center in O_1. Taking into account these relations, it is established that the fixed centroid is a circle with the center in O_1 and the radius R. Furthermore, the distance AI is equal, in any position, to the diameter of the support circle $2R$. So the distance $AI=2R$ and the mobile centroid is also a circle with the center in A and the radius $2R$.

For the analytical determination of the centroids, the fixed coordinate system $O_1x_1y_1$, the mobile coordinate system (solidly connected to the bar) Oxy and the angle θ as a time-varying parameter are chosen.

The parametric equations of the fixed centroid are:

$$\begin{cases} x_1 = IO_1 \cdot \cos 2\theta = R\cos 2\theta \\ y_1 = IO_1 \sin 2\theta = R\sin 2\theta \end{cases}; \tag{3.126}$$

from which eliminating the parameter θ we obtain the equation of the fixed centroid:

$$x_1^2 + y_1^2 = R^2, \tag{3.127}$$

i.e. a circle with center at O and radius R.

The parametric equations of the mobile centroid are:

$$\begin{cases} x = AB = 2R\cos\theta \\ y = AC = 2R\sin\theta \end{cases}. \tag{3.128}$$

Eliminating the parameter θ, we obtain:

$$x^2 + y^2 = 4R^2; \tag{3.129}$$

i.e. a circle with center O and radius $2R$.

To determine the angular velocity, we take into account that:

$$v_o = \omega \cdot IA = \omega \cdot 2R; \tag{3.130}$$

whence:

$$\omega = \frac{v_o}{2R} \quad \text{deci} \quad \varepsilon = 0. \tag{3.131}$$

The velocity of point B is determined with the relation:

$$v_B = \omega \cdot IB = \omega \cdot 2R \sin\alpha = \frac{\cdot v_o}{2R} \cdot 2R \sin\alpha = v_o \sin\alpha; \tag{3.132}$$

and has the direction after the bar AB.

To determine the acceleration, we write:

$$\begin{aligned}
\vec{a}_B &= \vec{a}_A + \vec{\varepsilon} x \overrightarrow{AB} - \omega^2 \overrightarrow{AB} \\
&= 0 + d\vec{k}x\left(2R\cos^2\theta\vec{i} + 2R\cos\theta\sin\theta\vec{j}\right) \\
&\quad - \omega^2\left(2R\cos^2\theta\vec{i} + 2R\cos\theta\sin\theta\vec{j}\right) \\
&= -(2\varepsilon R\cos\theta\sin\theta + 2R\omega^2\cos^2\theta)\vec{i} \\
&\quad + (2\varepsilon R\cos^2\theta - 2R\omega^2\cos\theta\sin\theta)\vec{j} \\
&= -\frac{v_o^2}{2R}\cos^2\theta\vec{i} - \frac{v_o^2}{2R}\cos\theta\sin\theta\vec{j}.
\end{aligned} \tag{3.133}$$

and one obtained the value: $a_B = \frac{v_o^2}{2R}\cos\theta$. Point A is the pole of accelerations.

A3. Determine the fixed and mobile centroid of the bar AB represented in Fig. 3.26 and having the length l.

Solution: Geometric solution: drawing perpendiculars to the velocities in A and B, we obtain point I (the instantaneous center of rotation). The quadrilateral O_1AIB is inscribable (two right angles) in the circle of diameter IO_1. During the movement of the bar, the angle AIB corresponds to the same segment AB, so the diameter O_1I of the circle is constant. So the mobile centroid is a circle of diameter O_1I:

$$O_1I = \frac{l}{\sin(\pi - \beta)} = \frac{l}{\sin\beta}\text{(sinus theorem)}. \tag{3.134}$$

The fixed centroid is a circle of radius O_1I, since, throughout the motion, point I remains at a constant distance O_1I from point O_1.

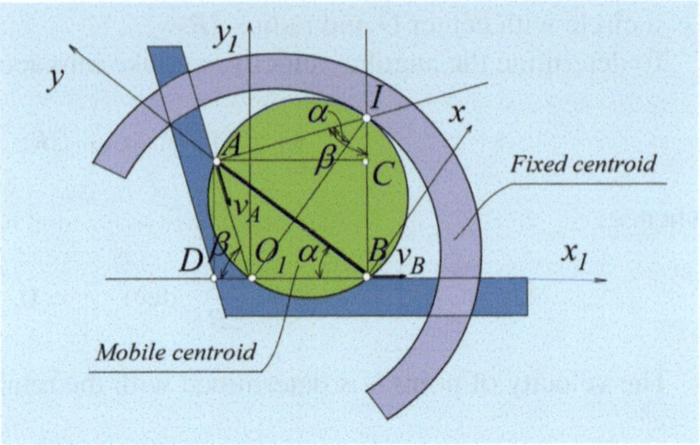

Fig. 3.26 Centroids of the bar

Semi-analytical solution:

The coordinates of the instantaneous center of rotation with respect to the two systems will be written:

$$x_1 = l\cos\alpha - O_1D\,;\quad O_1D = \frac{l\sin\alpha}{tg\beta}\,;\quad x_1 = l\cos\alpha - \frac{l\sin\alpha}{tg\beta}; \tag{3.135}$$

$$y_1 = l\sin\alpha + CI\,;\quad CI = \frac{l\cos\alpha}{tg\beta}\,;\quad y_1 = l\sin\alpha + \frac{l\cos\alpha}{tg\beta}. \tag{3.136}$$

Squaring these expressions and adding them term by term, we obtain the equation of the fixed centroid:

$$x_1^2 + y_1^2 = \frac{l^2}{tg^2\beta}\left(1 + tg^2\beta\right) = \frac{l^2}{\sin^2\beta}\,;\quad x_1^2 + y_1^2 = \frac{l^2}{tg^2\beta}. \tag{3.137}$$

So the fixed centroid is a circle of radius $\frac{l}{\sin\beta}$ with the center at the origin of the fixed axis system $O_1x_1y_1$.

The coordinates of the instantaneous center of rotation with respect to the mobile system are:

$$x = BI\cos\alpha\,;\quad y = BI\sin\beta; \tag{3.138}$$

where:

$$BI = BC_1 + C_1I = l\sin\alpha + \frac{l\cos\alpha}{tg\beta}; \tag{3.139}$$

$$x = l\sin\alpha\cos\alpha + \frac{l}{tg\beta}\cos^2\alpha = \frac{l}{2}\sin 2\alpha + \frac{l}{tg\beta}\left(\frac{1-\cos 2\alpha}{2}\right); \tag{3.140}$$

$$y = l\sin^2\alpha + \frac{l}{tg\beta}\sin\alpha\cos\alpha = \frac{l}{2}(1-\cos 2\alpha) + \frac{l}{tg\beta}\frac{\sin 2\alpha}{2}. \tag{3.141}$$

By squaring and adding the member-by-member relationships, a convenient grouping of the terms obtained can be achieved such that:

$$\left(x - \frac{l}{2tg\beta}\right)^2 + \left(y - \frac{l}{2}\right)^2 = \frac{l^2}{4} + \frac{l^2}{4tg^2\beta}, \tag{3.142}$$

or:

$$\left(x - \frac{l}{2tg\beta}\right)^2 + \left(y - \frac{l}{2}\right)^2 = \frac{l^2}{4\sin^2\beta}. \tag{3.143}$$

The mobile centroid is a circle of radius $\frac{l}{2\sin\beta}$ with center at point O and passing through points I, A, B and O_1. The velocity distribution for the bar sliding on the walls is given in Fig. 3.27.

A4. A bar moves in a plane so that end A always slides on the vertical wall, and the bar always passes through point B (Fig. 3.28). Determine the fixed and mobile centroids and represent the velocity distribution.

Solution: $AB = \frac{a}{\cos\alpha}$; $AI = \frac{AB}{\cos\alpha} = \frac{a}{\cos^2\alpha}$; $BI = AI\sin\alpha = \frac{a}{\cos^2\alpha}\sin\alpha$; $x_1 = BI\sin\alpha = atg^2\alpha$; $y_1 = -BI\cos\alpha = -atg\alpha$.

Equation of the fixed centroid is:

$$x_1 = \frac{1}{a}y_1^2.$$

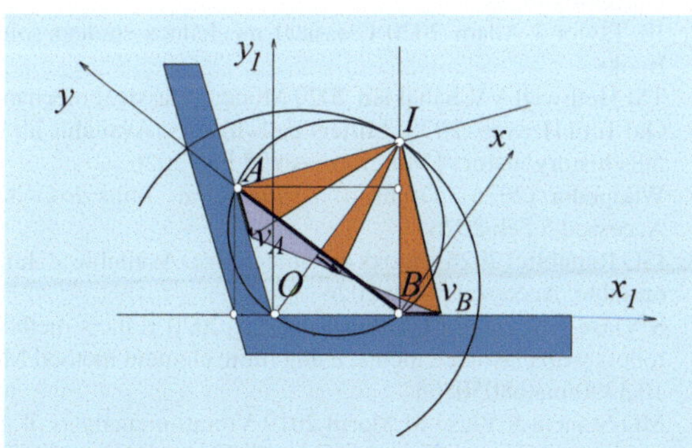

Fig. 3.27 Distribution of the velocities

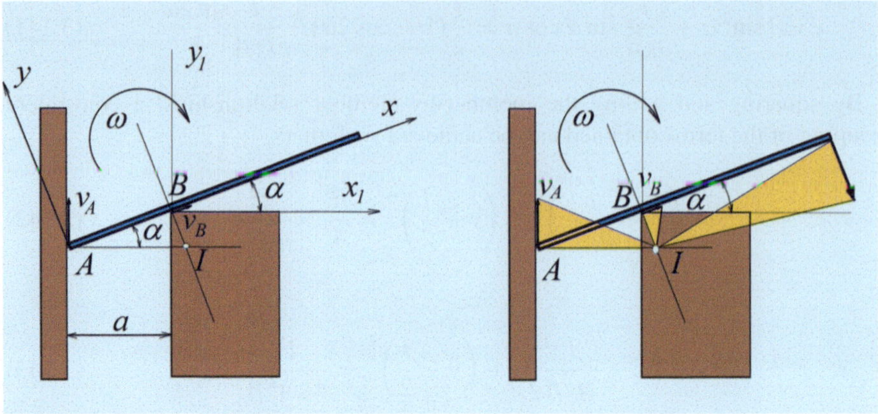

Fig. 3.28 The moving bar (left) and the distribution of the velocities (right)

Considering: $x = AB = \frac{a}{\cos\alpha}$; $y = -BI = -\frac{a\sin\alpha}{\cos^2\alpha}$, equation of the mobile centroid is:

$$y^2 = \frac{a^2\sin^2\alpha}{\cos^4\alpha} = \frac{a^2\left(1-\cos^2\alpha\right)}{\cos^4\alpha} = \frac{a^2\left(1-\frac{a^2}{x^2}\right)}{\frac{a^4}{x^4}} = \frac{x^2\left(x^2-a^2\right)}{a^2}.$$

References

1. D Morin 2008 Introduction to classical mechanics: with problems and solutions 1 Cambridge University Press
2. JR Taylor J Adam 2020 Classical mechanics student solutions manual University Science Books
3. TM Helliwell VV Sahakian 2020 Modern classical mechanics Cambridge University Press
4. Old Tool Heaven (2026) Millers Falls history. Available at: https://oldtoolheaven.com/millers-falls/history/history4.htm. Accessed 5 Feb 2026
5. Wikipedia (2026) Tîrbuşon. Available at: https://ro.wikipedia.org/wiki/Tirbu%C8%99on. Accessed 5 Feb 2026
6. ISO Republic (2026) Books on table image. Available at: https://isorepublic.com/photo/books-on-table. Accessed 5 Feb 2026
7. S Vlase I Negrean M Marin S Năstac 2020 Kane's method-based simulation and modeling robots with elastic elements, using finite element method Mathematics 8 5 805 https://doi.org/10.3390/math8050805
8. MD Stanciu S Vlase M Marin 2019 Vibration analysis of a guitar considered as symmetrical mechanical system Symmetry 11 6 727 https://doi.org/10.3390/sym11060727
9. ML Scutaru S Vlase M Marin 2023 Symmetrical mechanical system properties-based forced vibration analysis J Comput Appl Mech 54 4 501 514 https://doi.org/10.22059/JCAMECH.2023.366509.890

10. Marin, M (1996) Generalized solutions in elasticity of micropolar bodies with voids. Revista de la Academia Canaria de Ciencias 8(1):101–106. https://dialnet.unirioja.es/servlet/articulo?codigo=2899045
11. M Marin 1998 Contributions on uniqueness in thermoelastodynamics on bodies with voids Cienc Mat (Havana) 16 2 101 109
12. ML Scutaru S Vlase 2012 Some properties of motion equations describing the nonlinear dynamical response of a multibody system with flexible elements J Appl Math 1 628503https://doi.org/10.1155/2012/628503
13. Vlase S, Teodorescu PP (2013) Elasto-dynamics of a solid with a general 'rigid' motion using FEM model. Part I. Theoretical approach. Roman J Phys 58(7–8):872–881
14. S Vlase M Marin A Öchsner 2019 Considerations of the transverse vibration of a mechanical system with two identical bars Proc Inst Mech Eng Part L: J Mater: Des Appl 233 7 1318 1323
15. Stanciu E, Bencze A, Cazan A, Vlase S, Marin M, Öchsner A (2025) Effects of infill parameter variations on mechanical properties in 3D printed structures: comparative study with FEM analysis. J Comput Appl Mech 56(2):318–330. https://doi.org/10.22059/jcamech.2025.390563.1371
16. M Marin MM Bhatti OM Hapenciuc S Pirlog 2025 A spatial decay estimates for a thermoelastic Cosserat body without energy dissipation J Comput Appl Mech 56 3 611 626 https://doi.org/10.22059/jcamech.2025.396902.1506
17. Sharma K, Marin M, Kumar R (2025) Thermomechanical deformation in a micropolar thermoviscoelastic solid under the Moore-Gibson-Thompson heat equation with non-local and hyperbolic two-temperature effects. J Comput Appl Mech 56(4):720–736. https://doi.org/10.22059/jcamech.2025.397650.152

Chapter 4
Relative Motion

In the study of the motion of a body, we always need a referential to which we relate this motion. In current practice we can have situations where we know the motion of the body in relation to a reference system, but it is necessary to know the motion in relation to another reference system, which is in relative motion to the first one. The present chapter aims to solve precisely such situations. The following definitions are given:

Absolute motion is the movement of a body relative to a reference system considered fixed.

Relative motion is the movement with respect to the reference system considered mobile.

Transport motion is the movement of the mobile reference system relative to the fixed reference system.

Depending on the reference system to which the motion is related, we will talk about the trajectory, velocity and acceleration, absolute and relative of a point belonging of the body. Transport motion being the motion of a reference system with respect to another considered fixed system, is a motion of a rigid body that can be considered solidary with this reference system. Thus, it is characterized by the velocity and acceleration of the origin of the system and by the angular velocity and angular acceleration of the system in motion. If the body can be assimilated to a point, we are dealing with the relative movement of the material point. When the body cannot be assimilated to a point, we are dealing with the relative motion of the rigid body [2–4].

Figure 4.1 shows a crane bridge where the mentioned motion can be identified.

S. Vlase et al., *Models in Kinematics for Engineers*,
https://doi.org/10.1007/978-3-032-28113-5_4

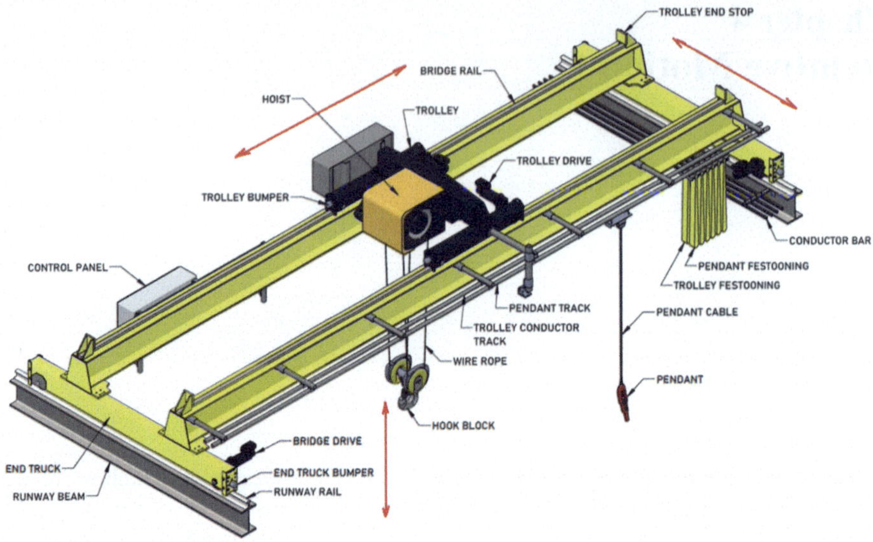

Fig. 4.1 Bridge crane [1]

4.1 The Relative Motion of the Material Point

4.1.1 The Absolute Velocity of the Material Point

Consider a fixed reference system $O_1x_1y_1z_1$ and a mobile system $Oxyz$ moving relative to the first one (Fig. 4.2). The velocity of the origin of the mobile reference system is $\vec{v}_0$, and the acceleration is $\vec{a}_0$. The system as a whole has an angular velocity $\vec{\omega}$ and an angular acceleration $\vec{\varepsilon}$. The material point M is defined by the vector $\overrightarrow{OM} = \vec{r}$ with the components related to the fixed reference system or the same vector $\overrightarrow{OM} = \vec{\rho}$ with the components related to the mobile reference system.

The connection between the components of the vector $\overrightarrow{OM}$ in the two reference systems is made by the rotation matrix $[R]$:

$$\{r\} = [R]\{\rho\} \tag{4.1}$$

The position of the material point in relation to the fixed reference system is given by:

$$\{r_M\} = \{r_0\} + \{r\} = \{r_0\} + [R]\{\rho\} \tag{4.2}$$

a relationship similar to that from the study of the movement of a point of the rigid body. The difference is that, if in the case of the rigid body, the OM size was constant,

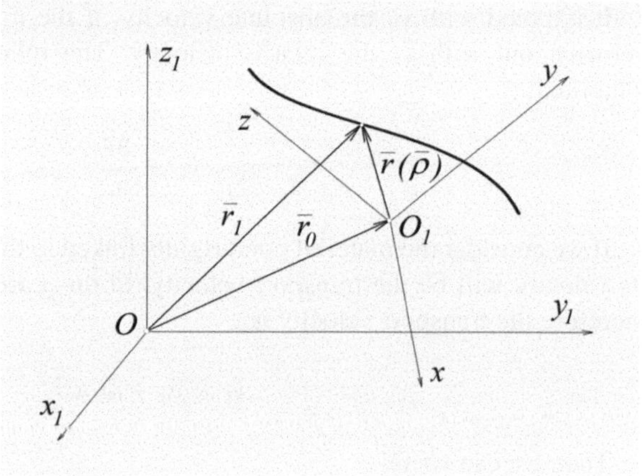

Fig. 4.2 Fixed and mobile reference system

in this case it varies. If we derive the written relation we will obtain:

$$\{\dot{r}_M\} = \{\dot{r}_0\} + [\dot{R}]\{\rho\} + [R]\{\dot{\rho}\} \tag{4.3}$$

In relation to the relationship obtained in the study of the rigid body, the expression $[R]\{\dot{\rho}\}$ appears in addition. The term $\{\dot{\rho}\}$ represents the variation of the vector $\overrightarrow{OM}$ in relation to the moving reference system, so it will be the relative speed of the material point. By multiplying by $[R]$ the vector is related to the fixed reference system. This velocity is denoted by $\frac{\partial \vec{r}}{\partial t}$. We mention that we are dealing only with a notation and not with a partial derivative. The other components appearing in the velocity expression were clarified before, for the motion of the rigid body. We therefore retain the vector $\vec{r}$ derivation formula:

$$\dot{\vec{r}}_M = \vec{\omega} \times \vec{r} + \frac{\partial \vec{r}}{\partial t} \tag{4.4}$$

The derivative of any vector $\vec{b}$ in motion with respect to the mobile reference system and with respect to the fixed reference system will have the same expression:

$$\dot{\vec{b}} = \vec{\omega} \times \vec{b} + \frac{\partial \vec{b}}{\partial t} \tag{4.5}$$

Then we will be able to write the expression of the absolute velocity of the material point in the form:

$$\vec{v}_M = \dot{\vec{r}}_M = \vec{v}_0 + \vec{\omega} \times \vec{r} + \frac{\partial \vec{r}}{\partial t} \tag{4.6}$$

It is noted with $\vec{v}_a$ the absolute velocity of the material point, $\vec{v}_t$ the transport velocity and with $\vec{v}_r$ the relative velocity. The relative velocity was previously identified:

$$\vec{v}_r = \frac{\partial \vec{r}}{\partial t} \tag{4.7}$$

If we consider the material point rigidly linked to the fixed reference system, then its velocity will be the transport velocity. In this case the term $[R]\{\dot{\rho}\}$ is zero and therefore the transport velocity is:

$$\vec{v}_t = \vec{v}_0 + \vec{\omega} \times \vec{r} \tag{4.8}$$

Then we can write:

$$\vec{v}_a = \vec{v}_t + \vec{v}_r \tag{4.9}$$

that is, the absolute velocity of the material point is obtained from the transport velocity and the relative velocity, vectorially composed.

4.1.2 Absolute Acceleration of the Material Point

To find the acceleration of the material point we derive the relationship that gives us the velocity:

$$\vec{a}_M = \ddot{\vec{r}}_M = \vec{a}_0 + \dot{\vec{\omega}} \times \vec{r} + \vec{\omega} \times \dot{\vec{r}} + \frac{d}{dt}\left(\frac{\partial \vec{r}}{\partial t}\right) \tag{4.10}$$

Taking into account the derivation relation presented previously, we can further write:

$$\vec{a}_M = \ddot{\vec{r}}_M = \vec{a}_0 + \dot{\vec{\omega}} \times \vec{r} + \vec{\omega} \times \left(\vec{\omega} \times \vec{r} + \frac{\partial \vec{r}}{\partial t}\right) + \vec{\omega} \times \frac{\partial \vec{r}}{\partial t} + \frac{\partial^2 \vec{r}}{\partial t^2} \tag{4.11}$$

or:

$$\vec{a}_M = \vec{a}_0 + \vec{\varepsilon} \times \vec{r} + \vec{\omega} \times (\vec{\omega} \times \vec{r}) + 2\vec{\omega} \times \frac{\partial \vec{r}}{\partial t} + \frac{\partial^2 \vec{r}}{\partial t^2} \tag{4.12}$$

To identify the acceleration of the transport movement, the vector $\overrightarrow{OM}$ is considered of constant length and then $\vec{v}_r = 0$. It results the transport acceleration of the material point:

$$\vec{a}_M = \vec{a}_t = \vec{a}_0 + \vec{\varepsilon} \times \vec{r} + \vec{\omega} \times (\vec{\omega} \times \vec{r}) \tag{4.13}$$

It becomes clear that the acceleration of the relative movement is obtained if we fix the moving reference system, that is, if we consider that at any moment of time we have $\omega = 0$, $\varepsilon = 0$. Putting these conditions, the relative acceleration of the material point will result from the expression of the acceleration:

$$\vec{a}_M = \vec{a}_r = \frac{\partial^2 \vec{r}}{\partial t^2} \tag{4.14}$$

It is found that the absolute acceleration of the material point is the vector sum of the transport acceleration, the relative acceleration and another term called complementary acceleration or Coriolis acceleration:

$$\vec{a}_C = 2\vec{\omega} \times \frac{\partial \vec{r}}{\partial t} = 2\vec{\omega} \times \vec{v}_r \tag{4.15}$$

Coriolis acceleration is a coupling term between transport motion and relative motion. It will express the superimposed influence of the rotational motion of the mobile reference system and the relative motion on the absolute acceleration.

So the expression of the absolute acceleration will be:

$$\vec{a}_a = \vec{a}_M = \vec{a}_t + \vec{a}_r + \vec{a}_C \tag{4.16}$$

Coriolis acceleration occurs when the mobile reference system performs a rotational movement. A very suggestive example is the case of the Earth, which performs the diurnal rotation around its axis.

Warm air masses from the equator rise, giving way to cold air masses that will come from the tropics. These, instead of moving in a longitudinal circle, will be diverted to the east. The prevailing winds in the Northern Hemisphere due to the warming of the air at the equator will be diverted to the West. In the South, they will also be diverted to the West, if they head towards the equator (Figs. 4.3 and 4.4).

The Coriolis acceleration, which gives rise to the Coriolis force, also explains the appearance of hurricanes (cyclones) large air masses with a rotational movement. The cold air that comes to replace the warm air in a geographical area acquires, thanks to the Coriolis force, a rotational movement. In the Northern Hemisphere this rotation will be anticlockwise, while in the Southern Hemisphere it will be clockwise (Figs. 4.5 and 4.6).

In Fig. 4.7, cyclones formed in the Northern Hemisphere can be observed, with a counter-clockwise movement. The equator line is highlighted by clouds formed by the condensation of vapors from the warm air that has risen.

Examples **E1**. A material point descends on the generator of a cone starting from its top with constant speed u. The cone rotates around the axis with constant angular velocity ω. Determine the absolute velocity and absolute acceleration of the material

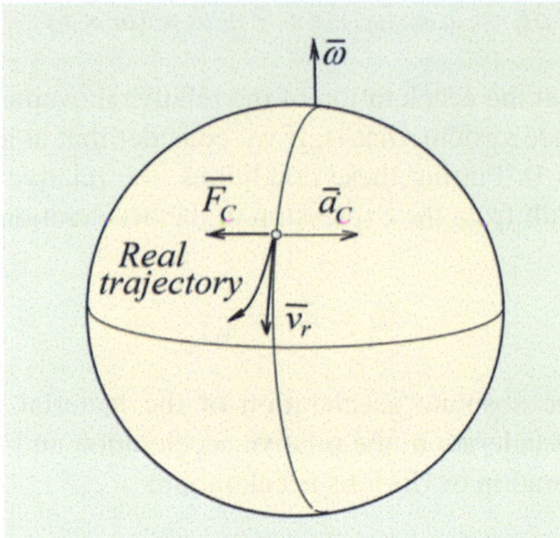

Fig. 4.3 Deviation due to the coriolis effect

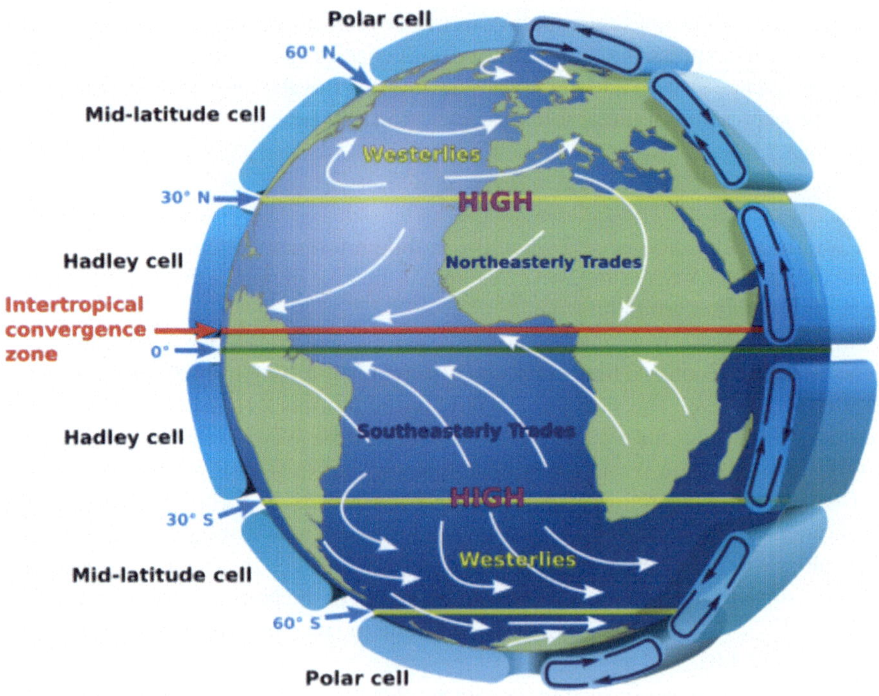

Fig. 4.4 Prevailing winds [5]

Fig. 4.5 Huricane Katrina[6]

Fig. 4.6 Huricane in Southern hemisphere [7]

Fig. 4.7 The direction of winds on earth [8]

point at a given moment. The height of the cone H and the angle at the top α are known (Fig. 4.8).

Answer: Analyzing Fig. 4.8, the relationships can be written:

$$\vec{r} = d\sin\alpha\vec{i} + (H - d\cos\alpha)\vec{k}_1 \; unde \; \vec{i} = \sin\theta\vec{i}_1 + \cos\theta\vec{j}_1, \tag{4.17}$$

so, it results:

$$\vec{r} = d\sin\alpha\sin\theta\vec{i}_1 + d\sin\alpha\cos\theta\vec{j}_1 + (H - d\cos\alpha)\vec{k}_1 \tag{4.18}$$

Further:

$$\begin{aligned}\vec{v}_r = \vec{u} &= u\sin\alpha\vec{i} - u\cos\alpha\vec{k}_1 \\ &= u\sin\alpha\sin\theta\vec{i}_1 + u\sin\alpha\cos\theta\vec{j}_1 - u\cos\alpha\vec{k}_1\end{aligned} \tag{4.19}$$

The angular velocity will be: $\vec{\omega} = \omega\vec{k}_1$, the angular acceleration $\varepsilon = 0$, and the relative acceleration $a_r = \dot{u} = 0$. The velocity and acceleration of the origin of the mobile system relative to the fixed reference system are zero. The absolute velocity and acceleration of the material point can now be easily obtained.

E2. A material point moves with constant velocity u along a chord of a circle at a distance s from the center. The circle rotates around the.

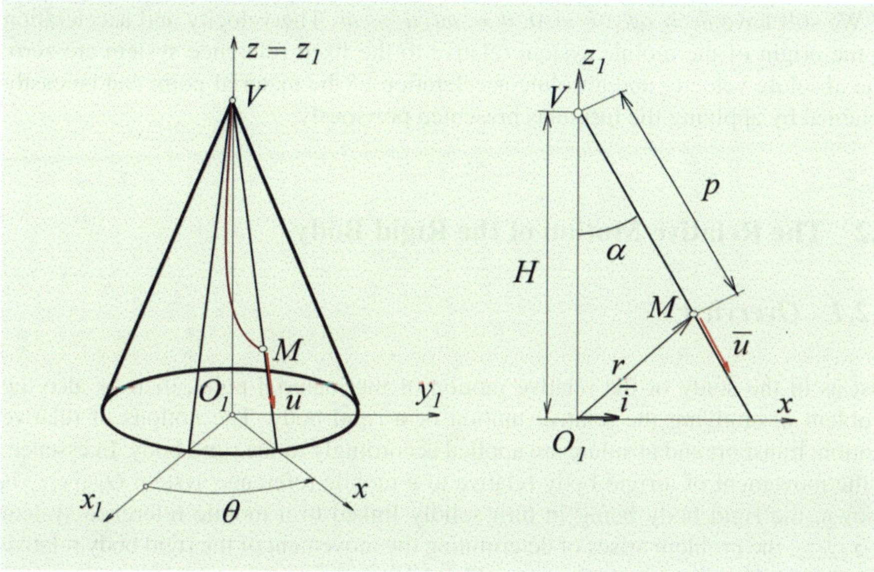

Fig. 4.8 A material point sliding on a cone

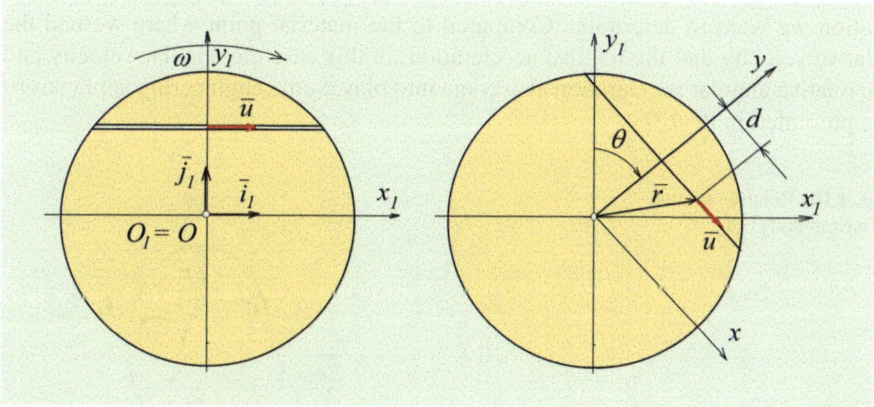

Fig. 4.9 A material point on a rotating disk

center with constant angular velocity ω Fig. 4.9. Determine the velocity and absolute acceleration of the material point at a given moment.

Answer. Following Fig. 4.9 it can be written:

$$\vec{r} = \vec{s} + \vec{d} = s\left(\sin\theta\vec{i}_1 + \cos\theta\vec{j}_1\right) + d\left(\cos\theta\vec{i}_1 + \sin\theta\vec{j}_1\right) \tag{4.20}$$

We still have $\vec{\omega} = \omega \vec{k}_1$, $\vec{\varepsilon} = 0$, $\theta = \omega t, d = ut$. The velocity and acceleration of the origin of the mobile system relative to the fixed reference system are zero. The absolute velocity and absolute acceleration of the material point can be easily obtained by applying the formulas presented previously.

4.2 The Relative Motion of the Rigid Body

4.2.1 Overview

Just as in the study of the relative motion of the material point, there is also the problem of studying the relative motion of a rigid body. The notions of relative motion, transport and absolute are applied accordingly to the rigid body. In essence, if the movement of a rigid body relative to a mobile reference system $O_1x_1y_1z_1$ is known, the rigid body being in turn solidly linked to a mobile reference system $O_2x_2y_2z_2$, the problem arises of determining the movement of the rigid body relative to a fixed reference system $O_0x_0y_0z_0$ (Fig. 4.10).

For ease of writing, the three reference systems will be denoted by T_1, T_2, and T_0, respectively. Knowing the motion of a rigid body comes back to knowing the motion of any of its points. We choose an arbitrary point of the rigid body whose motion we want to determine. Compared to the material point where we had the relative velocity and the relative acceleration, in this case the angular velocity and the relative angular acceleration also come into play. Some engineering applications are presented in [9–17].

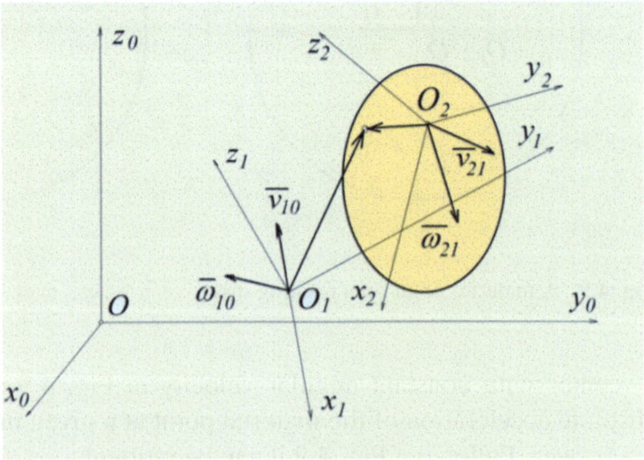

Fig. 4.10 Relative motion of a rigid body

4.2.2 The Study of Velocities

The position vector of the material point relative to T_2 is denoted by $\vec{r}_2$, and the position vector relative to T_1 is denoted by $\vec{r}_1$. Taking into account the previously stated, we have:

- the relative velocity of point M will be the velocity of M with respect to the moving trihedral T_1. If we consider T_1 fixed, then the speed of point M relative to it is given by the relationship known from rigid body kinematics:

$$\vec{v}_r = \vec{v}_{21}^M = \vec{v}_{21} + \vec{\omega}_{21} \times \vec{r}_2 \tag{4.21}$$

where $\vec{v}_{21}$ represents the velocity of point O_2 relative to O_1.

- the transport velocity of point M is the speed that M would have if the rigid body were solidly connected to T_1. In this case, the position vector of M with respect to T_1 being $\vec{r}_1$, we obtain:

$$\vec{v}_t = \vec{v}_{10} + \vec{\omega}_{10} \times \vec{r}_1 \tag{4.22}$$

- the absolute velocityis obtained by summing the transport and relative velocity:

$$\vec{v}_a = \vec{v}_t + \vec{v}_r \tag{4.23}$$

from which the final formula of the velocity of point M with respect to the fixed reference system T_0 will result:

$$\vec{v}_{20}^M = \vec{v}_{10} + \vec{v}_{21} + \vec{\omega}_{10} \times \vec{r}_1 + \vec{\omega}_{21} \times \vec{r}_2 \tag{4.24}$$

The notation for trihedral, respectively T_0, T_1, T_2, was made to be able to generalize the previously obtained relationship. Thus, if we have "n" trihedrals T_0, T_1,…T_n, the rigid body being solidly connected to the last trihedral T_n, we aim to determine the absolute velocity of a point M of the rigid, depending on the relative movements of one trihedral to another. It is denoted by $\vec{r}_0, \vec{r}_1, ...\vec{r}_n$ the position vectors of the point M with respect to the respective trihedral, with $\vec{v}_{ij}$ the velocity of the origin of the trihedral "i" with respect to "j", with $\vec{\omega}_{ij}$ the angular velocity of the trihedral "i" with respect to "j".

It results then the velocity of point M relative to the fixed reference system given by the relation:

$$\vec{v}_{n0}^M = \vec{v}_{10} + \vec{v}_{21} + \cdots + \vec{v}_{n,n-1} + \vec{\omega}_{10} \times \vec{r}_1 + \vec{\omega}_{21} \times \vec{r}_2 + \cdots + \vec{\omega}_{n,n-1} \times \vec{r}_n \tag{4.25}$$

or:

$$\vec{v}_{n0}^{M} = \sum_{i=1}^{n} \vec{v}_{i,i-1} + \sum_{i=1}^{n} \vec{\omega}_{i,i-1} \times \vec{r}_i \tag{4.26}$$

We further propose to express the absolute angular velocity of the last reference frame in relation to the fixed reference frame, by means of the relative angular velocities of one reference frame in relation to another. If we consider two points of the rigid body M and N, then Poisson's relation gives us:

$$\vec{v}_{n0}^{N} = \vec{v}_{n0}^{M} + \vec{\omega}_{n0} \times \overrightarrow{MN} \tag{4.27}$$

With the previous relation that expresses the absolute velocities of points M and N by means of relative movements, we obtain:

$$\vec{v}_{n0}^{M} = \sum_{i=1}^{n} \vec{v}_{i,i-1} + \sum_{i=1}^{n} \vec{\omega}_{i,i-1} \times \overrightarrow{O_iM} \tag{4.28}$$

$$\vec{v}_{n0}^{N} = \sum_{i=1}^{n} \vec{v}_{i,i-1} + \sum_{i=1}^{n} \vec{\omega}_{i,i-1} \times \overrightarrow{O_iN} \tag{4.29}$$

Subtraction results in:

$$\vec{v}_{n0}^{N} - \vec{v}_{n0}^{M} = \sum_{i=1}^{n} \vec{\omega}_{i,i-1} \times \left(\overrightarrow{O_iN} - \overrightarrow{O_iM} \right) = \left(\sum_{i=1}^{n} \vec{\omega}_{i,i-1} \right) \times \overrightarrow{MN} \tag{4.30}$$

Comparing the relationship obtained with Poisson's formula results:

$$\vec{\omega}_{n0} = \sum_{i=1}^{n} \vec{\omega}_{i,i-1} \tag{4.31}$$

4.2.3 The Study of Accelerations

It is operated in the same way as in the case of calculating the relative velocity. The motion of the trihedral T_2 with respect to T_1 is characterized by the vectors $\vec{a}_{21}$ (the acceleration of O_2 with respect to O_l), $\vec{v}_{21}$, $\vec{\varepsilon}_{21}$ (the angular acceleration of T_2 with respect to T_1) and $\vec{\omega}_{21}$, and the motion of the trihedral T_1 with respect to T_0 is characterized by the vectors $\vec{a}_{10}$ (the acceleration of O_1 with respect to O_0), $\vec{v}_{10}$, $\vec{\varepsilon}_{10}$ (angular acceleration of T_1 with respect to T_0) and $\vec{\omega}_{10}$. With the introduced notations we have:

- the relative acceleration of point M will be the acceleration of M with respect to the mobile trihedral T_1. If we consider T_1 fixed, then the acceleration of M relative to it is given by the relationship known from rigid body kinematics:

$$\vec{a}_r = \vec{a}_{21}^M = \vec{a}_{21} + \vec{\varepsilon}_{21} \times \vec{r}_2 + \vec{\omega}_{21} \times (\vec{\omega}_{21} \times \vec{r}_2) \tag{4.32}$$

- the transport acceleration of point M is the acceleratia that M would have if the rigid body were solidly connected to T_1. In this case it is obtained:

$$\vec{a}_t = \vec{a}_{10}^M = \vec{a}_{10} + \vec{\varepsilon}_{10} \times \vec{r}_1 + \vec{\omega}_{10} \times (\vec{\omega}_{10} \times \vec{r}_1) \tag{4.33}$$

- the relative velocity will be:

$$\vec{v}_r = \vec{v}_{21} + \vec{\omega}_{21} \times \vec{r}_2 \tag{4.34}$$

- the Coriolis acceleration is calculated with the relation:

$$\vec{a}_C = 2\vec{\omega}_t \times \vec{v}_r = 2\vec{\omega}_{10} \times (\vec{v}_{21} + \vec{\omega}_{21} \times \vec{r}_2) \tag{4.35}$$

The absolute acceleration is obtained by summing the three accelerations:

$$\vec{a}_a = \vec{a}_r + \vec{a}_t + \vec{a}_C \tag{4.36}$$

from which the final formula of the acceleration of point M with respect to the fixed reference system T_0 will result:

$$\begin{aligned} \vec{a}_{20}^M &= \vec{a}_{10} + \vec{a}_{21} + \vec{\varepsilon}_{10} \times \vec{r}_1 + \vec{\varepsilon}_{21} \times \vec{r}_2 + \vec{\omega}_{10} \times (\vec{\omega}_{10} \times \vec{r}_1) \\ &+ \vec{\omega}_{21} \times (\vec{\omega}_{21} \times \vec{r}_2) + 2\vec{\omega}_{10} \times (\vec{v}_{21} + \vec{\omega}_{21} \times \vec{r}_2). \\ &+ 2\vec{\omega}_{10} \times (\vec{v}_{21} + \vec{\omega}_{21} \times \vec{r}_2). \end{aligned} \tag{4.37}$$

The previously obtained relation can be generalized. Thus, if we have "n" trihedrals $T_0, T_1, \ldots T_n$, the mobile one being solidly connected to the last trihedral T_n, we aim to determine the absolute acceleration of a point M of the rigid according to the relative movements of one trihedral to another.

It is noted with $\vec{r}_i$ the position vectors of point M relative to the respective trihedral, with $\vec{a}_{ij}$ the acceleration of the origin of the trihedral "i" relative to "j", with $\vec{\varepsilon}_{ij}$ the angular acceleration of the trihedral "i" relative to "j". It is calculated for the beginning $\vec{\varepsilon}_{n0}$. It is obtained:

$$\vec{\varepsilon}_{n0} = \frac{d\vec{\omega}_{n0}}{dt} = \sum_{i=1}^{n} \frac{d\vec{\omega}_{i,i-1}}{dt} \tag{4.38}$$

Vectors $\vec{\omega}_{i,i-1}$ are defined relative to moving reference systems, so the derivation relations established at the beginning of the chapter apply. It results:

$$\vec{\varepsilon}_{n0} = \sum_{i=1}^{n} \frac{\partial \vec{\omega}_{i,i-1}}{\partial t} + \sum_{i=1}^{n} \left(\vec{\omega}_{i-1,0} \times \vec{\omega}_{i,i-1}\right) \tag{4.39}$$

The angular velocity $\vec{\omega}_{i-1,0}$ represents the angular transport velocity of the $T_{i\text{-}1}$ trihedral, and $\frac{\partial \vec{\omega}_{i,i-1}}{\partial t} = \vec{\varepsilon}_{i,i-1}$. Taking into account the relationship $\vec{\omega}_{i-1,0} = \sum_{j=1}^{n} \vec{\omega}_{j,j-1}$, previously determined, we obtain:

$$\vec{\varepsilon}_{n0} = \sum_{i=1}^{n} \vec{\varepsilon}_{i,i-1} + \sum_{i=1}^{n} \left(\vec{\omega}_{i-1,0} \times \vec{\omega}_{i,i-1}\right) + \sum_{i=1}^{n} \sum_{j=1}^{i-1} \left(\vec{\omega}_{i-1,0} \times \vec{\omega}_{j,j-1}\right) \tag{4.40}$$

We calculate $\vec{a}_{n0}^{M}$ by determining the accelerations from close to close. Thus, for the $\vec{a}_{i0}^{M}$ calculated function of the movement with respect to the T_{i-1} trihedral, we obtain:

- transport acceleration:

$$\vec{a}_t = \vec{a}_{i-1,0}^{M} \tag{4.41}$$

- relative acceleration:

$$\vec{a}_r = \vec{a}_{i,i-1} + \vec{\varepsilon}_{i,i-1} \times \vec{r}_i + \vec{\omega}_{i,i-1} \times \left(\vec{\omega}_{i,i-1} \times \vec{r}_i\right) \tag{4.42}$$

- Coriolis acceleration:

$$\vec{a}_C = 2\vec{\omega}_{i-1,0} \times \left(\vec{v}_{i,i-1} + \vec{\omega}_{i,i-1} \times \vec{r}_i\right) \tag{4.43}$$

Summing the relations one obtains the results:

$$\begin{aligned} \vec{a}_a = \vec{a}_{i,0}^{M} = \vec{a}_{i-1,0}^{M} = \vec{a}_{i,i-1} + \vec{\varepsilon}_{i,i-1} \times \vec{r}_i + \vec{\omega}_{i,i-1} \times \left(\vec{\omega}_{i,i-1} \times \vec{r}_i\right) \\ +2\vec{\omega}_{i-1,0} \times \left(\vec{v}_{i,i-1} + \vec{\omega}_{i,i-1} \times \vec{r}_i\right) \end{aligned} \tag{4.44}$$

Writing these relations for $i = 1,2,\ldots n$, adding them and reducing similar terms, we finally get:

$$\vec{a}^{M}_{n,0} = \sum_{i=1}^{n} \vec{a}_{i,i-1} + \sum_{i=1}^{n} \vec{\varepsilon}_{i,i-1} \times \vec{r}_i + \sum_{i=1}^{n} \left[\vec{\omega}_{i,i-1} \times \left(\vec{\omega}_{i,i-1} \times \vec{r}_i\right)\right] + \\ +2\sum_{i=1}^{n}\sum_{j=1}^{i-1} \vec{\omega}_{i-1,0} \times \left(\vec{v}_{i,i-1} + \vec{\omega}_{j,j-1} \times \vec{r}_i\right) \tag{4.45}$$

Analogously to the calculation of velocities, the acceleration of a point N can be obtained when considering M as the origin:

$$\vec{a}^{N}_{n0} = \vec{a}^{M}_{n0} + \vec{\varepsilon}_{n0} \times \overrightarrow{MN} + \vec{\omega}_{n0} \times \left(\vec{\omega}_{n0} \times \overrightarrow{MN}\right) \tag{4.46}$$

Applications:

A3.1. A rectangular frame ($BC = OD = R$) rotates with $\omega =$ const. around the axis of bearings AF generating a cylinder (Fig. 4.10). At the same time on the side CD a point M falls freely with acceleration g. The velocity and acceleration of point M at a given moment are required.

Solution: The fixed system is the frame with bearings A and F. The mobile system is the frame that performs a rotational motion around the axis AF. The relative motion is the rectilinear motion of M on CD. The transport motion is performed by M immobilized on the frame, that is, a circular motion with radius R and angular velocity ω. The absolute motion is the motion of M relative to the frame (Fig. 4.11)

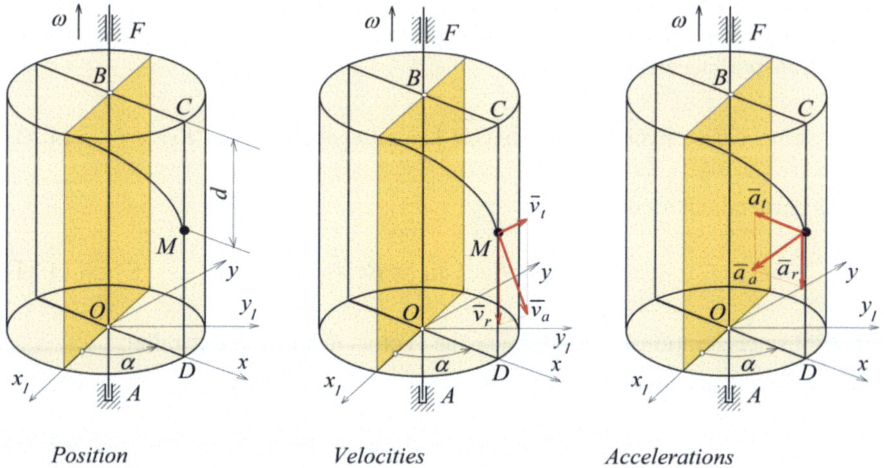

Fig. 4.11 Cylinder in an uniform rotation

Study of velocities: Relative velocity is the free fall velocity:

$$v_r = gt \tag{4.47}$$

and as a vector:

$$\vec{v}_r = -gt\vec{k} \tag{4.48}$$

Transport velocity is in the mobile coordinate system

$$\vec{v}_t = \vec{\omega}x\overrightarrow{OM} = \vec{\omega}x\left[\left(H - d)\vec{k} + R\vec{i}\right)\right] = \omega R\vec{j} \tag{4.49}$$

with the value:

$$v_t = R\omega \tag{4.50}$$

The absolute velocity is:

$$\overline{v}_a = \overline{v}_r + \overline{v}_t = -gt\vec{k} + \omega R\vec{j} \tag{4.51}$$

and as $\vec{k}$ (v_r) and $\vec{j}$ (v_t) are perpendicular:

$$v_a = |\overline{v}_a| = \sqrt{v_r^2 + v_t^2} = \sqrt{R^2\omega^2 + g^2t^2} \tag{4.52}$$

and is contained in a plane tangent to the cylinder in M.

Study of accelerations: Relative acceleration is $\vec{a}_r = -g\vec{k}$ with the value $a_r = g$. The transport acceleration is:

$$\vec{a}_t = \vec{\varepsilon}x\overrightarrow{OM} + \vec{\omega}x\left(\vec{\omega}x\overrightarrow{OM}\right) = \omega\vec{k}x\left(\omega R\vec{j}\right) = -\omega^2R\vec{i} \tag{4.53}$$

and has perpendicular components:

$$\vec{a}_t\left(a_{tn} = R\omega^2;\, a_{t\tau} = R\varepsilon = 0\right) \tag{4.54}$$

Coriolis acceleration is zero because the vectors $\overline{\omega}$ and $\underset{r}{\vec{v}}$ are parallel.

The absolute acceleration is:

$$\overline{a}_a = \overline{a}_r + \overline{a}_t + \overline{a}_c = -g\vec{k} - R\omega^2\vec{i} \tag{4.55}$$

and is contained in a plane of the spiral. The acceleration modulus is:

$$a_a = |a_a| = \sqrt{a_r^2 + a_{tn}^2} = \sqrt{g^2 + R^2\omega^4}. \tag{4.56}$$

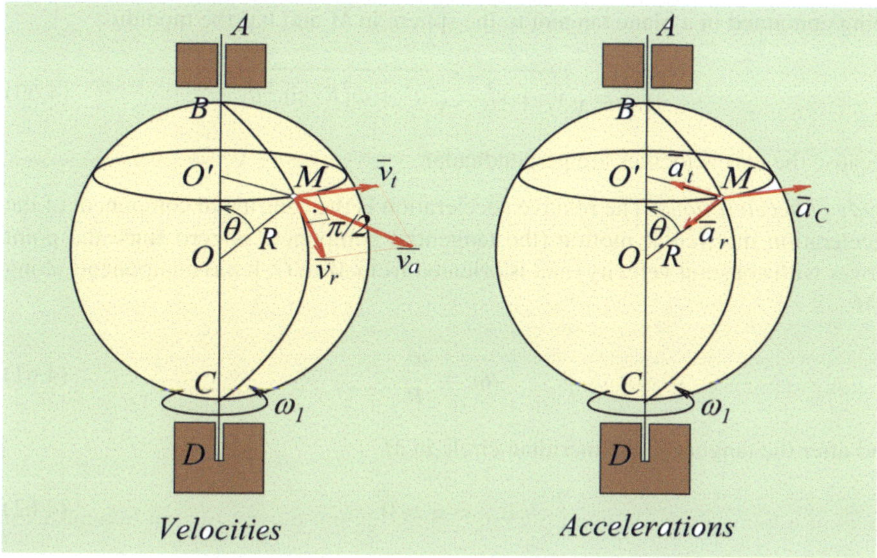

Fig. 4.12 Velocities and acceleration on the surface of a sphere

A3.2 A frame *ABCD*, which has a semicircular portion *BC* of radius *R*, rotates with ω_1 = const. around the axis of the bearings *AD*, generating a sphere. At the same time, a point *M* moves on the semicircular portion of the frame, with a speed u = const. along the meridian (Fig. 4.12). The speed and acceleration of point *M* are required.

Solution: The fixed reference system is the frame with bearings *A* and *D*. The mobile reference system is the frame, which performs a rotational motion with an angular velocity ω_1 around the axis *AD*. The relative motion is the circular motion of *M* with a speed u on a circle of radius $OM = R$ (on the frame). The transport motion is the motion of *M* fixed on the frame, i.e. on a circle of radius the center being $O'M = R\sin\theta = R\sin\frac{u}{R}t$, the center beeing O' with an angular velocity ω_1. The absolute motion is the motion of *M* relative to the frame.

Study of velocities: The relative velocity is:

$$v_r = u; \vec{v}_r = -u\sin\theta\vec{k} + u\cos\theta\vec{i} \tag{4.57}$$

The transport velocity is:

$$v_t = O'M\,\omega_1 = \omega_1 R\sin\theta \quad v_t = \omega_1 R\sin\theta\vec{j} \tag{4.58}$$

The absolute velocity is (Fig. 4.12):

$$\overline{v}_a = \overline{v}_r + \overline{v}_t = -u\sin\theta\vec{k} + u\cos\theta\vec{i} + \omega_1 R\sin\theta\vec{j} \tag{4.59}$$

being contained in a plane tangent to the sphere in M and has the modulus:

$$|\overline{v}_a| = \sqrt{v_r^2 + v_t^2} = \sqrt{u^2 + \omega_1^2 R^2 \sin^2\theta}, \tag{4.60}$$

because the two velocities are perpendicular.

Study of accelerations: The relative acceleration is the centripetal component of the acceleration in circular motion (the tangential component is zero since the point moves with constant velocity) and is oriented from M to O. It has components along OM:

$$a_{rn} = \frac{u^2}{R} \tag{4.61}$$

and after the tangent to the meridian circle in M

$$a_{r\tau} = \dot{u} = 0 \tag{4.62}$$

so:

$$\vec{a}_r\left(a_{rn} = \frac{u^2}{R};\, a_{r\tau} = \dot{u} = 0\right) \tag{4.63}$$

or, vectorial:

$$\vec{a}_r = -\frac{u^2}{R}\sin\theta\vec{i} - \frac{u^2}{R}\sin\theta\vec{k} \tag{4.64}$$

The transport acceleration has a centripetal component, along the direction of the circle radius MO',

$$a_{tn} = O'M\,\omega_1^2 = \omega_1^2 R\sin\theta \tag{4.65}$$

and a tangential component

$$a_{t\tau} = O'M\,\varepsilon_1 = 0. \tag{4.66}$$

It results:

$$\vec{a}_t\left(a_{tn} = O'M \cdot \omega_1^2 = \omega_1^2 R\sin\theta;\, a_{t\tau} = O'M \cdot \varepsilon_1 = 0\right), \tag{4.67}$$

or vectorial:

$$\vec{a}_t = -\omega_1^2 R\sin\theta\vec{i} \tag{4.68}$$

The Coriolis acceleration is

$$2\overline{\omega}_t \times \overline{v}_r = 2\overline{\omega}_1 \times \overline{v}_r \tag{4.69}$$

and has the support perpendicular to the plane of the frame, the direction in Fig. 4.12 and the expression:

$$\vec{a}_C = 2\omega_1 \vec{k} x \left(-u \sin\theta \vec{k} + u \cos\theta \vec{i} \right) = 2\omega_1 u \cos\theta \vec{j} \tag{4.70}$$

The modulus is

$$a_C = 2\omega_t v_r \sin(\overline{\omega}_t, \overline{v}_r) = 2\omega_1 \frac{u}{R} R \sin\left(\theta + \frac{\pi}{2}\right) = 2\omega_1 u \cos\theta.$$

$$a_C = 2\omega_t v_r \sin(\overline{\omega}_t, \overline{v}_r) = 2\omega_1 \frac{u}{R} R \sin\left(\theta + \frac{\pi}{2}\right) = 2\omega_1 u \cos\theta. \tag{4.72}$$

The absolute acceleration is

$$\overline{a}_a = \overline{a}_r + \overline{a}_t + \overline{a}_c \tag{4.73}$$

and has the modulus

$$\begin{aligned} a_a &= \sqrt{\overline{a}_r^2 + \overline{a}_t^2 + \overline{a}_c^2 + 2\overline{a}_r \cdot \overline{a}_t} \\ &= \sqrt{\frac{u^4}{R^2} + \omega_1^4 R^2 \sin^2\theta + 4\omega_1^2 u^2 \cos^2\theta + 2\omega_1^2 u^2 \sin^2\theta} \end{aligned} \tag{4.74}$$

because:

$$\overline{a}_r \cdot \overline{a}_c = \overline{a}_t \cdot \overline{a}_c = 0. \tag{4.75}$$

A3.3 The arc of a circle AB of radius R rotates around the axis Oz with a constant angular velocity ω. A material point M moves on the disk from A to B with a constant velocity in the modulus and equal to u. It is required to determine the velocity and absolute acceleration of the point M in an arbitrary position defined by the angle α, if the arc AB is located at that moment in the plane yOz; it will be specified for $\alpha = 0$ and $\alpha = \pi/2$ (Figs. 4.12 and 4.13).

Answer:

$$v_{a(M)} = \sqrt{\omega^2 R^2 (1 - \cos\alpha)^2 + u^2}$$

$$a_{a(M)} = \sqrt{4\omega^2 u^2 \sin^2\alpha + \left[\frac{u^2}{R} \cos\alpha - \omega^2 R (1 - \cos\alpha) \right]^2 + \frac{u^4}{R^2} \sin^2\alpha}$$

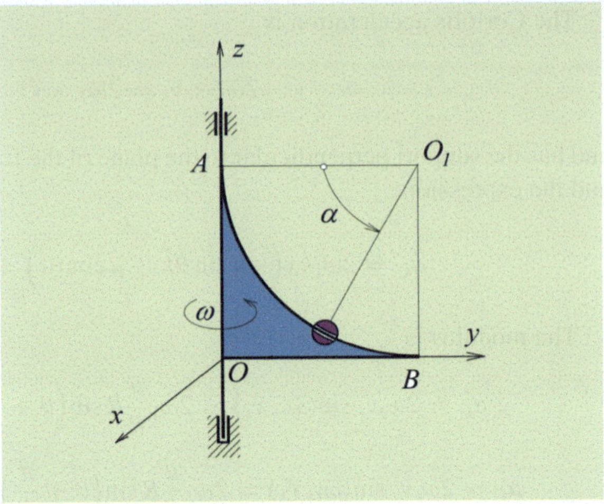

Fig. 4.13 Relative motion of a point

So, for $\alpha = 0$, $v_{a(M)} = u$; $a_{a(M)} = u^2/R$; and for $\alpha = \pi/2$, $v_{a(M)} = \sqrt{\omega^2 R^2 + u^2}$; $a_{a(M)} = \sqrt{4\omega^2 u^2 + \omega^4 R^2 + u^4/R^2}$.

References

1. TS Overhead Crane (2026) Overhead bridge cranes. Accessed 5 Feb 2026. Available at: https://www.tsoverheadcrane.com/overhead-bridge-cranes
2. Morin D (2008) Introduction to classical mechanics: with problems and solutions, 1st edn. Cambridge University Press
3. Taylor JR, Adam J (2020) Classical mechanics student solutions manual. University Science Books
4. Helliwell TM, Sahakian VV (2020) Modern classical mechanics. Cambridge University Press
5. Wikipedia (2026) Prevailing winds. Accessed 5 Feb 2026. Available at: https://en.wikipedia.org/wiki/Prevailing_winds
6. Wikipedia (2026) Hurricane Katrina—image. Accessed 5 Feb 2026. Available at: https://en.wikipedia.org/wiki/Hurricane_Katrina#/media/File%3AKatrina_2005-08-28_1700Z.jpg
7. Hypothetical Hurricanes Fandom (2026) 2017–18 Southern hemisphere cyclone season—image. Accessed 5 Feb 2026. Available at: https://hypotheticalhurricanes.fandom.com/wiki/2017-18_Southern_Hemisphere_Cyclone_Season?file=Marcus_2018-03-21_1505Z.png
8. Wikipedia (2026) NASA blue marble—image. Accessed 5 Feb 2026. Available at: https://ro.m.wikipedia.org/wiki/Fi%C8%99ier:Nasa_blue_marble.jpg
9. Vlase S, Negrean I, Marin M, Năstac S (2020) Kane's method-based simulation and modeling robots with elastic elements, using finite element method. Mathematics 8(5):805. https://doi.org/10.3390/math8050805
10. Stanciu MD, Vlase S, Marin M (2019) Vibration analysis of a guitar considered as symmetrical mechanical system. Symmetry 11(6):727. https://doi.org/10.3390/sym11060727
11. Scutaru ML, Vlase S, Marin M (2023) Symmetrical mechanical system properties-based forced vibration analysis. J Comput Appl Mech 54(4):501–514. https://doi.org/10.22059/JCAMECH.2023.366509.890

12. Scutaru ML, Vlase S (2012) Some properties of motion equations describing the nonlinear dynamical response of a multibody system with flexible elements. J Appl Math 1:628503. https://doi.org/10.1155/2012/628503
13. Vlase S, Teodorescu PP (2013) Elasto-dynamics of a solid with a general 'rigid' motion using FEM model. Part I. Theoretical approach. Romanian J Phys 58(7–8):872–881
14. Vlase S, Marin M, Öchsner A (2019) Considerations of the transverse vibration of a mechanical system with two identical bars. Proc Inst Mech Eng Part L: J Mater: Des Appl 233(7):1318–1323
15. Stanciu E, Bencze A, Cazan A, Vlase S, Marin M, Öchsner A (2025) Effects of infill parameter variations on mechanical properties in 3D printed structures: comparative study with FEM analysis. J Comput Appl Mech 56(2):318–330. https://jcamech.ut.ac.ir/article_102096.html
16. Marin M, Bhatti MM, Hapenciuc OM, Pirlog S (2025) A spatial decay estimates for a thermoelastic Cosserat body without energy dissipation. J Comput Appl Mech 56(3):611–626
17. Ailawalia P, Marin M, Priyanka (2025) Moore-Gibson-Thompson thermoelastic medium with variable thermal conductivity. J Comput Appl Mech 56(4):791–801

Chapter 5
Kinematics of Multibody Systems

5.1 Constraints. Kinematic Conditions

5.1.1 Constraints

Links represent kinematic restrictions imposed on the material point, material point systems, rigid body or rigid body systems. For example, a material point must move so that it is always on a given curve. A connection can generally be expressed by a geometric relation of the form [1–3]:

$$f(x, y, z, \dot{x}, \dot{y}, \dot{z}, t) = 0; \tag{5.1}$$

or:

$$f(q_1, q_2, \ldots, q_n, \dot{q}_1, \dot{q}_2, \ldots, \dot{q}_n, t) = 0, \tag{5.2}$$

in the case of generalized coordinates. Constraints can be classified as follows:

- *scleronomic*, if they do not depend on time;
- *rheonomic*, if they depend on time.

Another classification of constraints is in:

- *holonomic*, when the derivatives of the coordinates with respect to time do not appear explicitly;
- *nonholonomic*, when the derivatives of the coordinates appear in the mathematical expression of the constraints.

The study of constraints will be resumed in the analytical mechanics chapter. In this chapter, the most used kinematic constraints (liaisons), defined as kinematic couples, will be analyzed.

S. Vlase et al., *Models in Kinematics for Engineers*,
https://doi.org/10.1007/978-3-032-28113-5_5

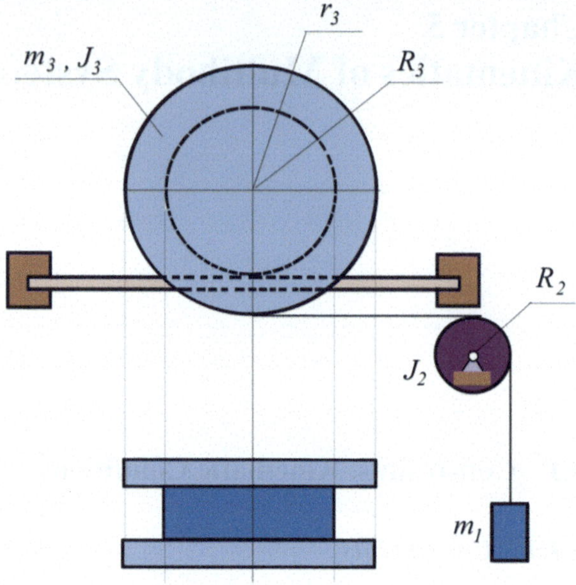

Fig. 5.1 System of bodies

In the following we present the kinematic conditions imposed by the links for some simple systems are presented as an example.

Example 1 Consider a system of connected bodies as in Fig. 5.1. The roller 3 rolls on a horizontal plane being pulled by a weight 1 that passes over a roller 2. We propose to see what are the kinematic conditions imposed by this system of connected bodies. There are three modes of motion (physical possibility of motion) for the system (Fig. 5.2):

(a) Pure rolling. In this case, the roller 3 will roll, without slipping, on the horizontal plane and taking into account the wire that ensures the connection between the bodies and that has the same linear speed along it at a given moment, it will be possible to write the kinematic conditions, which link the speeds:

$$\mathrm{v}_1 = \omega_2 R_2; \quad \mathrm{v}_1 = \omega_3(\mathrm{R}_3 - \mathrm{r}_3); \quad \mathrm{v}_3 = \omega_3 \mathrm{r}_3; \tag{5.3}$$

or, in compact form:

$$\begin{Bmatrix} v_1 \\ \omega_2 \\ \omega_3 \\ v_3 \end{Bmatrix} = \begin{Bmatrix} 1 \\ \frac{1}{R_2} \\ \frac{1}{R_3 - r_3} \\ \frac{r_3}{R_3 - r_3} \end{Bmatrix} v_1. \tag{5.4}$$

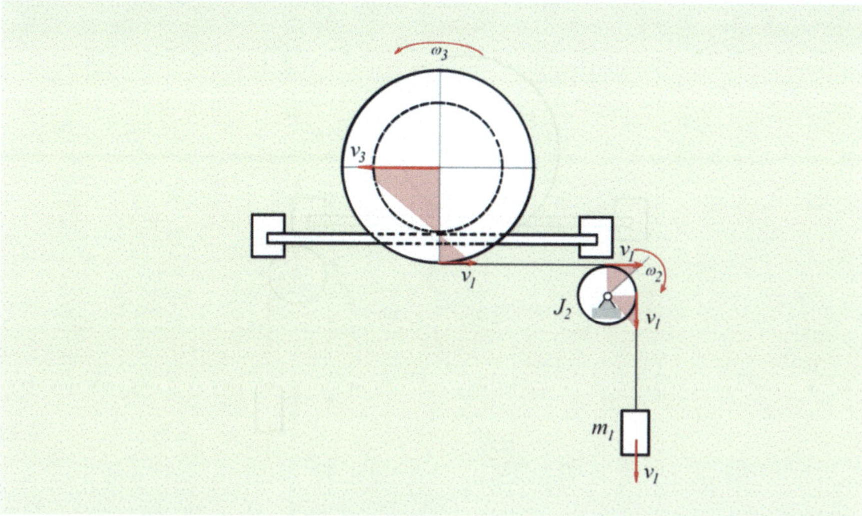

Fig. 5.2 Kinematic liaisons

With v_1 was noted the speed of the mass m_1, with ω_2 the angular speed of the cylinder 2, with v_3 the speed of the center of mass of the roller 3 and with ω_3 its angular speed. If the written relations are derived with respect to time, the relations between the accelerations will be obtained:

$$\begin{Bmatrix} a_1 \\ \varepsilon_2 \\ \varepsilon_3 \\ a_3 \end{Bmatrix} = \begin{Bmatrix} 1 \\ \frac{1}{R_2} \\ \frac{1}{R_3-r_3} \\ \frac{r_3}{R_3-r_3} \end{Bmatrix} a_1, \tag{5.5}$$

where the notations are obvious. This case is the most common in practice, in current applications.

(b) Rolling and sliding (Fig. 5.3). In this case, body 3 will roll and at the same time slide along the bar. The system will have two degrees of freedom.

In this case the kinematic conditions will be:

$$v_1 = \omega_2 R_2;\ v_1 - u_1 = \omega_3(R_3 - r_3);\ v_3 + u = \omega_3 r_3. \tag{5.6}$$

Doing the calculations we get:

$$\omega_2 = \frac{v_1}{R_2};\ \omega_3 = \frac{v_1}{R_3 - r_3} - \frac{\mathrm{u}}{R_3 - r};\ v_3 = v_1\frac{r_3}{R_3 - r_3} - u\frac{R_3}{R_3 - r_3}; \tag{5.7}$$

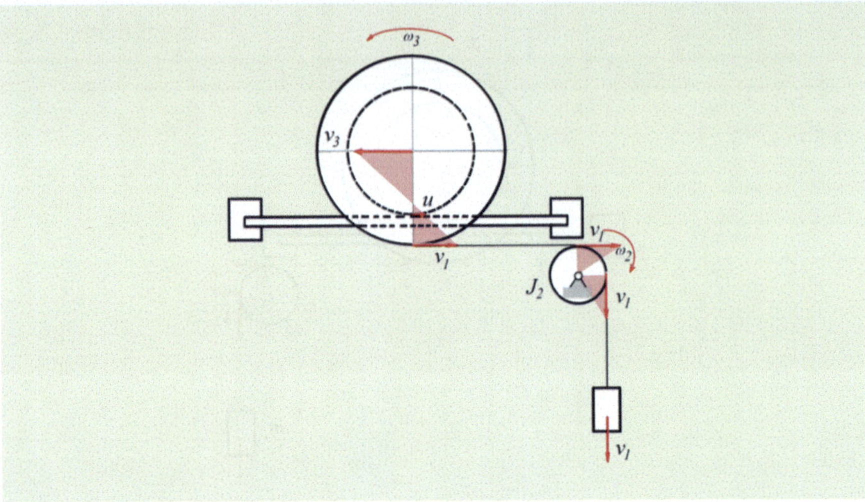

Fig. 5.3 Rolling with sliding

or:

$$\begin{Bmatrix} v_1 \\ \omega_2 \\ \omega_3 \\ v_3 \end{Bmatrix} = \begin{bmatrix} 1 & 0 \\ \frac{1}{R_2} & 0 \\ \frac{1}{R_3 - r_3} & -\frac{1}{R_3 - r_3} \\ \frac{r_3}{R_3 - r_3} & -\frac{R_3}{R_3 - r_3} \end{bmatrix} \begin{Bmatrix} v_1 \\ \dot{u} \end{Bmatrix}. \tag{5.8}$$

Deriving the written relationship with respect to time yields:

$$\begin{Bmatrix} a_1 \\ \varepsilon_2 \\ \varepsilon_3 \\ a_3 \end{Bmatrix} = \begin{bmatrix} 1 & 0 \\ \frac{1}{R_2} & 0 \\ \frac{1}{R_3 - r_3} & -\frac{1}{R_3 - r_3} \\ \frac{r_3}{R_3 - r_3} & -\frac{R_3}{R_3 - r_3} \end{bmatrix} \begin{Bmatrix} a_1 \\ \ddot{u} \end{Bmatrix}. \tag{5.9}$$

(c) *Pure sliding* (Fig. 5.4)

Theoretically, there could also be a mode of movement that consists of the translation of the mosorel along the plane.

This way can only be done in particular cases, practically being a rarely encountered situation.

In this case the kinematic conditions will be:

$$v_1 = \omega_2 R_2; \quad v_1 = v_3; \quad \omega_3 = 0. \tag{5.10}$$

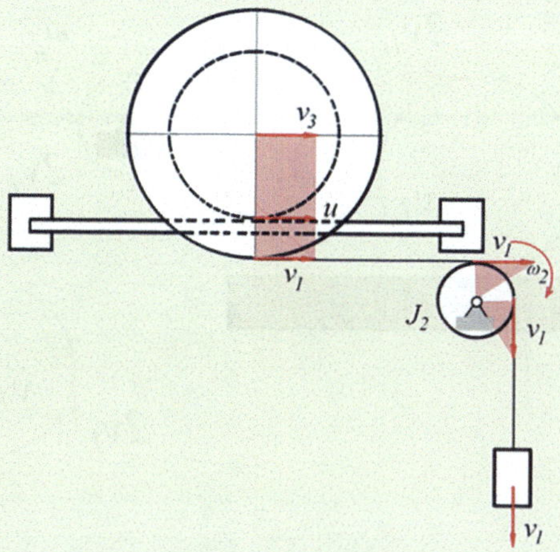

Fig. 5.4 Pure sliding

where from:

$$\omega_2 = \frac{v_1}{R_2}, \tag{5.11}$$

or:

$$\begin{Bmatrix} v_1 \\ \omega_2 \\ v_3 \end{Bmatrix} = \begin{bmatrix} 1 \\ \frac{1}{R_2} \\ 1 \end{bmatrix} v_1. \tag{5.12}$$

Deriving the written relationship with respect to time yields:

$$\begin{Bmatrix} a_1 \\ \varepsilon_2 \\ a_3 \end{Bmatrix} = \begin{bmatrix} 1 \\ \frac{1}{R_2} \\ 1 \end{bmatrix} a_1. \tag{5.13}$$

Example 2 Give the cylinders system in Fig. 5.5. We intend to determine the constraints between the kinematic parameters that describe the motion of each cylinder.

(a) Pure rolling. In this case, the cylinder 3 will roll, without slipping, on the horizontal plane and taking into account the wire that ensures the connection between the bodies and that has the same linear speed along it at a given moment, it will be possible to write the kinematic conditions as:

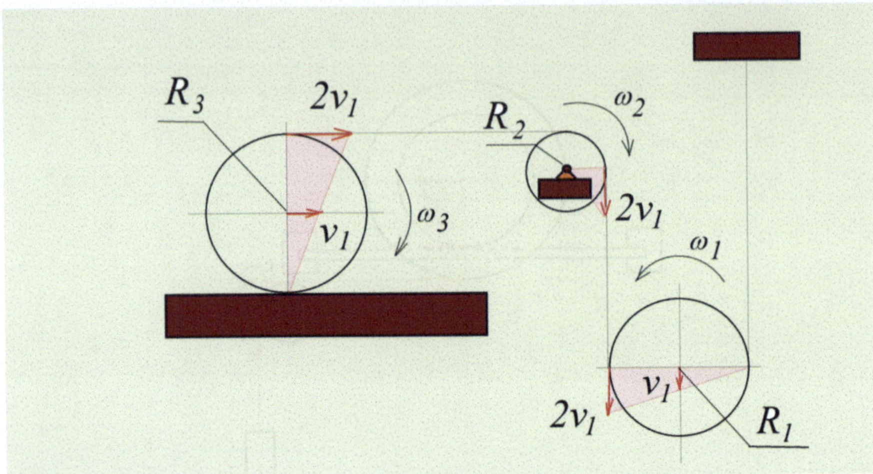

Fig. 5.5 Kinematic liaisons

$$\mathrm{v}_1 = \omega_1 \mathrm{R}_1;\quad 2\mathrm{v}_1 = \omega_2 \mathrm{R}_2;\quad 2\mathrm{v}_1 = \omega_3 2\mathrm{R}_3;\quad \mathrm{v}_3 = \omega_3 \mathrm{R}_3; \tag{5.14}$$

or:

$$\left\{\begin{matrix} \mathrm{v}_1 \\ \omega_1 \\ \omega_2 \\ \omega_3 \\ \mathrm{v}_3 \end{matrix}\right\} = \left\{\begin{matrix} 1 \\ \frac{1}{R_1} \\ \frac{2}{\mathrm{R}_2} \\ \frac{1}{\mathrm{R}_3} \\ 1 \end{matrix}\right\} v_1 . \tag{5.15}$$

With v_1 was noted the velocity of the mass m_1, with ω_1 the angular velocity of cylinder 1, with ω_2 the angular velocity of cylinder 2, with v_3 the velocity of the center of mass of cylinder 3 and with ω_3 the angular velocity of it. If the written relations are derived with respect to time, the liaisons between the accelerations will be obtained:

$$\left\{\begin{matrix} \mathrm{a}_1 \\ \varepsilon_1 \\ \varepsilon_2 \\ \varepsilon_3 \\ \mathrm{a}_3 \end{matrix}\right\} = \left\{\begin{matrix} 1 \\ \frac{1}{R_1} \\ \frac{2}{\mathrm{R}_2} \\ \frac{1}{\mathrm{R}_3} \\ 1 \end{matrix}\right\} a_1 , \tag{5.16}$$

where the notations are obvious.

(b) *Rolling and sliding* (Fig. 5.6). In this case, body 3 will roll and at the same time slide along the horizontal plane.

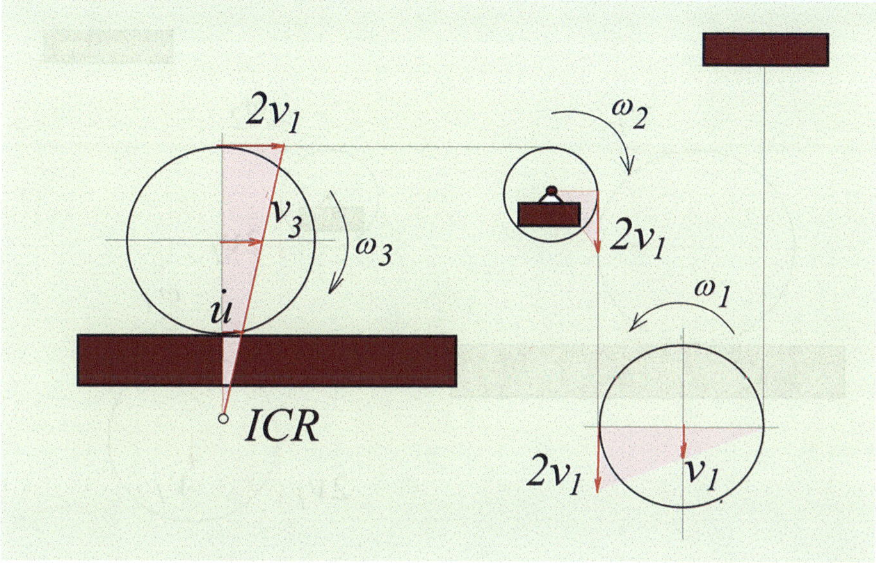

Fig. 5.6 Rolling and sliding

The kinematic conditions will be:

$$\mathrm{v}_1 = \omega_1 R_1; \quad 2v_1 = \omega_2 R_2; \quad 2v_1 - \dot{u} = 2\omega_3 R_3; \quad v_3 - \dot{u} = \omega_3 R_3. \tag{5.17}$$

Doing the calculations we get:

$$\begin{Bmatrix} \mathrm{v}_1 \\ \omega_1 \\ \omega_2 \\ \omega_3 \\ v_3 \end{Bmatrix} = \begin{bmatrix} 1 & 0 \\ \frac{1}{\mathrm{R}_1} & 0 \\ \frac{2}{\mathrm{R}_2} & 0 \\ \frac{1}{\mathrm{R}_3} & -\frac{1}{2\mathrm{R}_3} \\ 1 & -\frac{1}{2} \end{bmatrix} \begin{Bmatrix} v_1 \\ \dot{u} \end{Bmatrix}. \tag{5.18}$$

Deriving the written relationship with respect to time yields:

$$\begin{Bmatrix} \mathrm{a}_1 \\ \varepsilon_1 \\ \varepsilon_2 \\ \varepsilon_3 \\ a_3 \end{Bmatrix} = \begin{bmatrix} 1 & 0 \\ \frac{1}{\mathrm{R}_1} & 0 \\ \frac{2}{\mathrm{R}_2} & 0 \\ \frac{1}{\mathrm{R}_3} & -\frac{1}{2R_3} \\ 1 & -\frac{1}{2} \end{bmatrix} \begin{Bmatrix} a_1 \\ \ddot{u} \end{Bmatrix}. \tag{5.19}$$

(c) *Pure sliding*. It is a rare case (Fig. 5.7) as several geometric and mass conditions must be met for this mode of motion to exist.

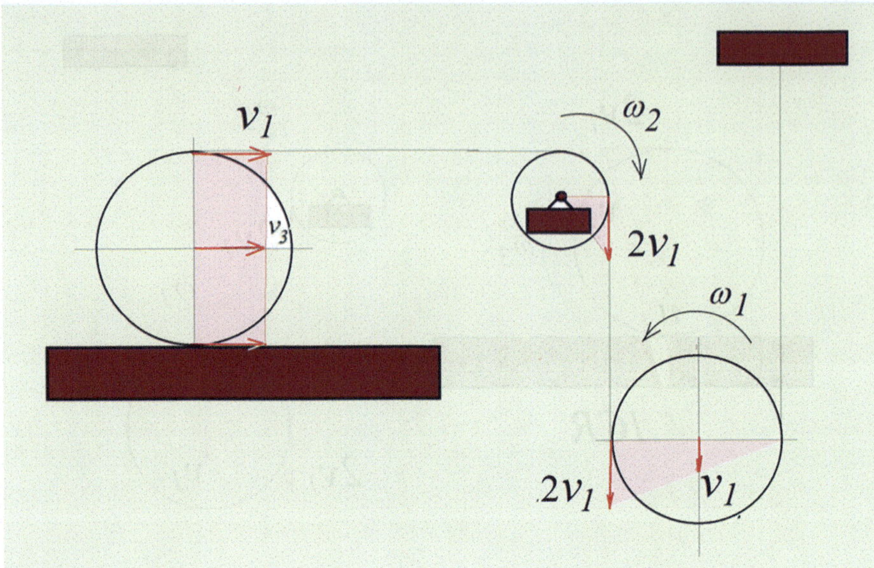

Fig. 5.7 Pure sliding

$$\mathrm{v}_1 = \omega_1 \mathrm{R}_1; \quad 2\mathrm{v}_1 = \omega_2 \mathrm{R}_2; \quad 2\mathrm{v}_1 = v_3, \tag{5.20}$$

or:

$$\begin{Bmatrix} v_1 \\ \omega_1 \\ \omega_2 \\ v_3 \end{Bmatrix} = \begin{Bmatrix} 1 \\ \frac{1}{R_1} \\ \frac{2}{R_2} \\ 2 \end{Bmatrix} v_1. \tag{5.21}$$

If the written relations are derived with respect to time, the connections between the accelerations will be obtained:

$$\begin{Bmatrix} a_1 \\ \varepsilon_1 \\ \varepsilon_2 \\ a_3 \end{Bmatrix} = \begin{Bmatrix} 1 \\ \frac{1}{R_1} \\ \frac{2}{R_2} \\ 2 \end{Bmatrix} a_1; \tag{5.22}$$

where the notations are obvious.

Example 3 There is a system made up of two cylinders, where the motion is made thanks to wires as in Fig. 5.8. Weight 1 sets the system in motion. The problem is to determine the links between the kinematic parameters that ensure the movement of the system.

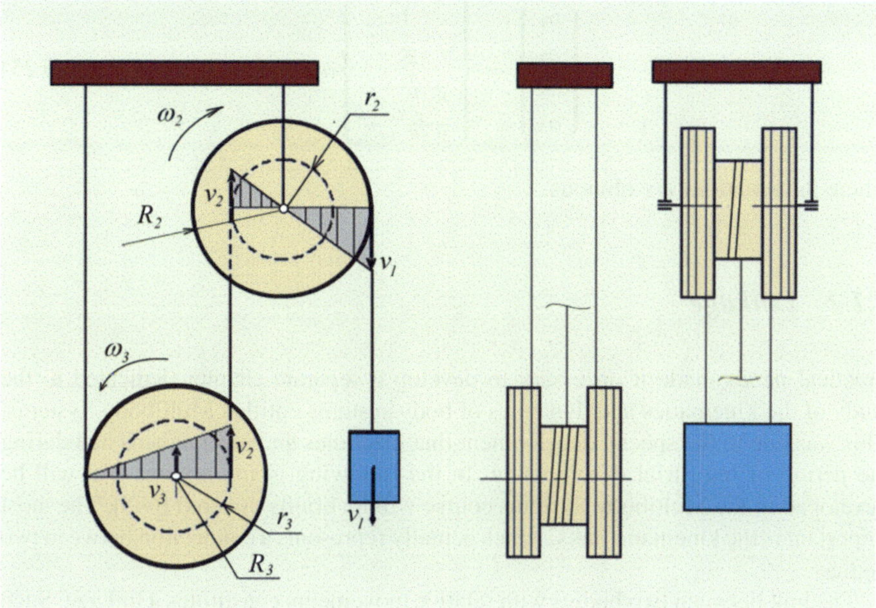

Fig. 5.8 A system with three bodies

If we take into account the wires that ensure the connections between the bodies, the relationships between the linear and angular velocities of the different bodies can be written:

$$v_1 = \omega_2 R_2; \quad \omega_2 \mathrm{r}_2 = \omega_3(R_3 + r_3); \quad v_3 = \omega_3 R_3; \tag{5.23}$$

or:

$$\begin{Bmatrix} v_1 \\ \omega_2 \\ \omega_3 \\ v_3 \end{Bmatrix} = \begin{Bmatrix} 1 \\ \frac{1}{R_2} \\ \frac{\mathrm{r}_2}{\mathrm{R}_2(R_3+r_3)} \\ \frac{\mathrm{r}_2 R_3}{\mathrm{R}_2(R_3+r_3)} \end{Bmatrix} v_1. \tag{5.24}$$

With v_1 was noted the velocity of the mass m_1, with ω_2 the angular velocity of the cylinder 2, with v_3 the velocity of the center of mass of the cylinder 3 and with ω_3 the angular velocity of it. If the written relations are derived with respect to time, the relations between the accelerations will be obtained,

$$\begin{Bmatrix} a_1 \\ \varepsilon_2 \\ \varepsilon_3 \\ a_3 \end{Bmatrix} = \begin{Bmatrix} 1 \\ \frac{1}{R_2} \\ \frac{r_2}{R_2(R_3+r_3)} \\ \frac{r_2 R_3}{R_2(R_3+r_3)} \end{Bmatrix} a_1, \tag{5.25}$$

where the notations are obvious.

5.1.2 Linkage

Practical needs made it necessary to develop a separate chapter dedicated to the study of the kinematics and dynamics of body systems entitled Multibody Systems. This was due to the special development that machines and mechanisms had during the period of industrial development. In the following, some notions that will be developed in the Multibody Systems course will be briefly covered [6–8]. The most important is the kinematic links, which actually represents a connection between two bodies.

The link between two bodies with relative movements constitutes a linkage. Such a constraint decreases the number of degrees of freedom of the body. Figures 5.9, 5.10, 5.11, 5.12, 5.13, 5.14, 5.15, 5.16 and 5.17 show a series of kinematic couples and the degrees of freedom allowed by them. In essence, the problem is as follows: a body has 6 degrees of freedom with respect to another body. If there are connections between the two bodies, the number of degrees of freedom will decrease. Linkages represent several types of connections that are frequently made in technical applications. The shape of these linkage by the links we want to ensure but also by the technological capacity to achieve them. By preparing a library of linkage, algorithms can be developed to describe the movement of bodies.

The notion of a kinematic loop refers to a set of elements linked together by linkage and is currently replaced by the notion of a multibody system.

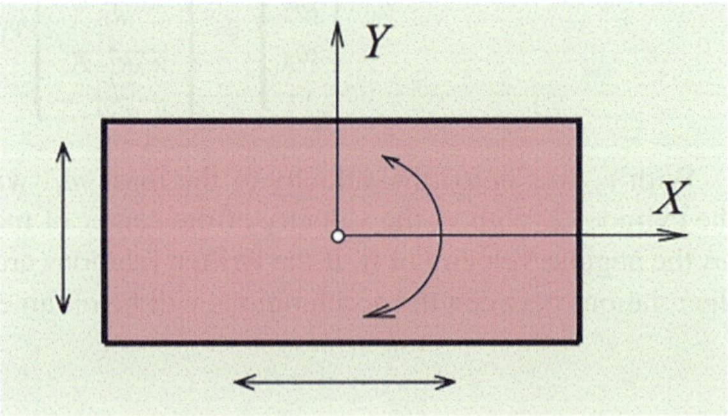

Fig. 5.9 The degrees of freedom of a rigid body with parallel plane motion

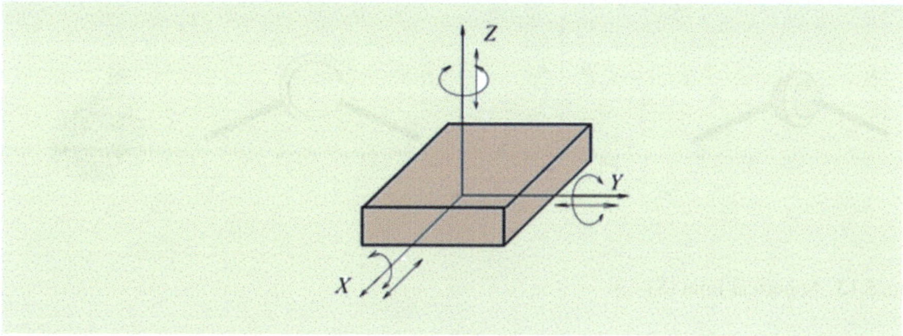

Fig. 5.10 The degrees of freedom of a rigid with general motion

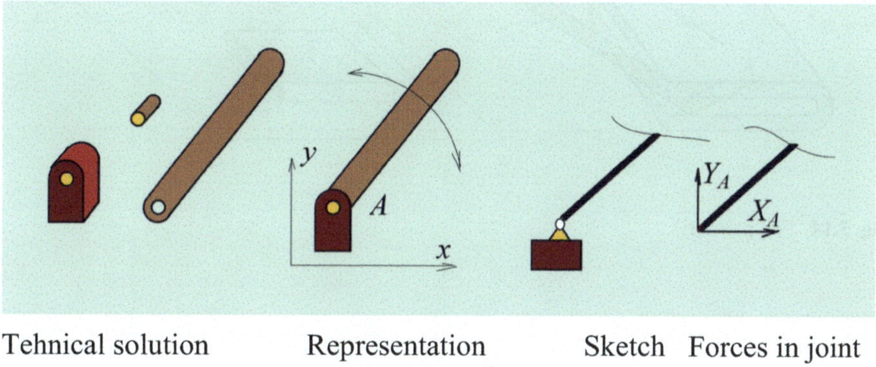

Fig. 5.11 Cylindrical joint

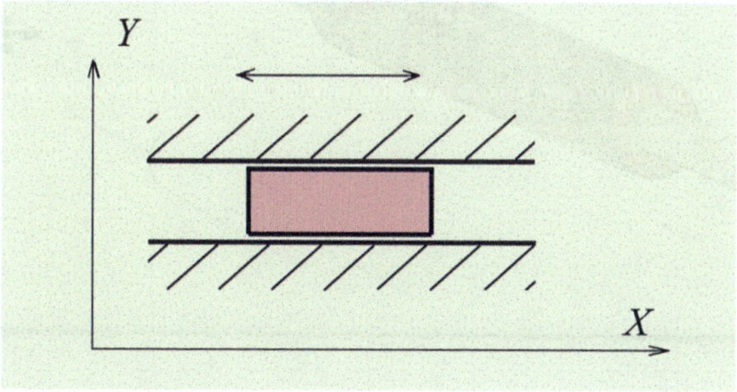

Fig. 5.12 Translation linkage (slider)

The notion of a mechanism, defined as a closed kinematic loop and studied as such for a long period, is considered too restrictive at the moment and is included in that of a multibody system.

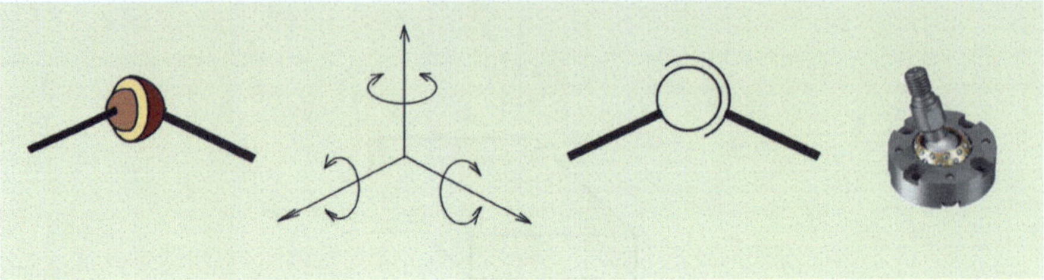

Fig. 5.13 Spherical joint (S)

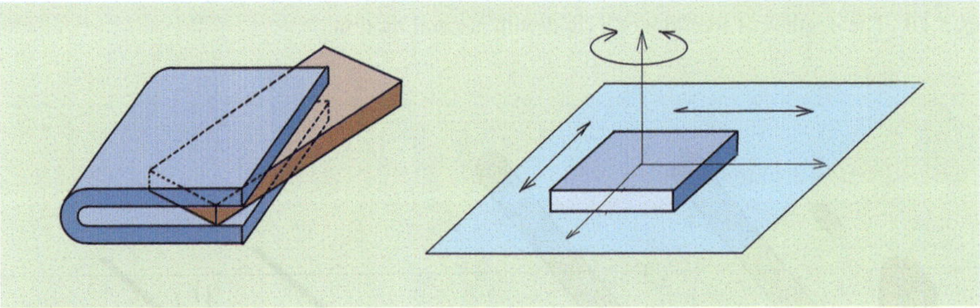

Fig. 5.14 Flat pairs (E)

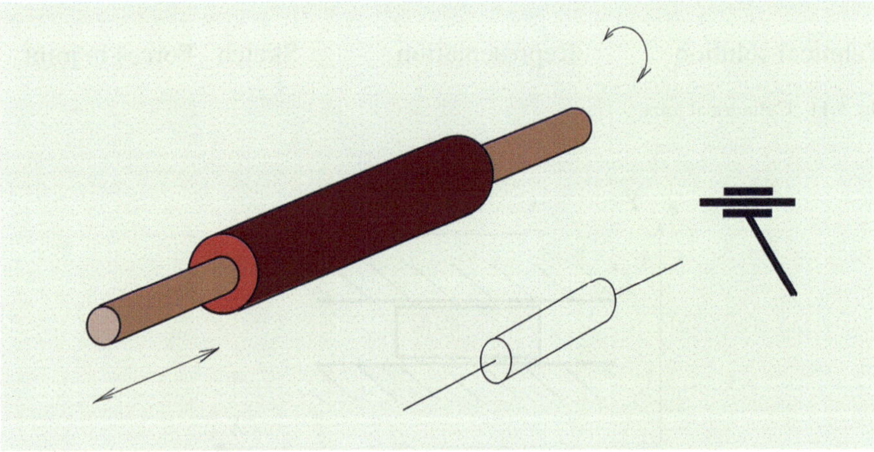

Fig. 5.15 Cylindrical linkage (C)

The theory of mechanisms, developed somewhat artificially, refers to a small number of situations that can appear in mechanics but which, at a given moment, were distinguished by the extent of practical applications. For this reason, the possibility of the theoretical development of the discipline, in parallel with Mechanics, was

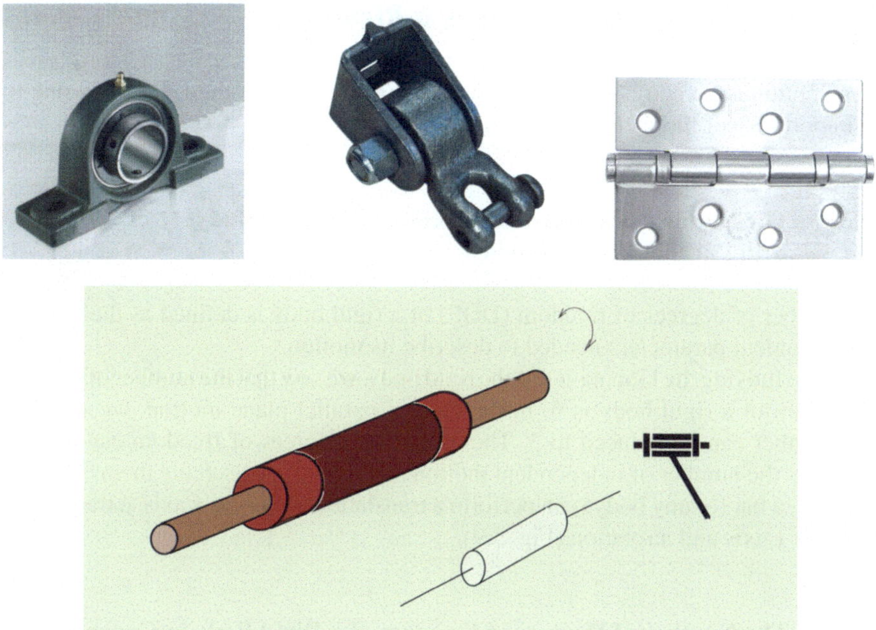

Fig. 5.16 Revolute linkage/hinge (R)

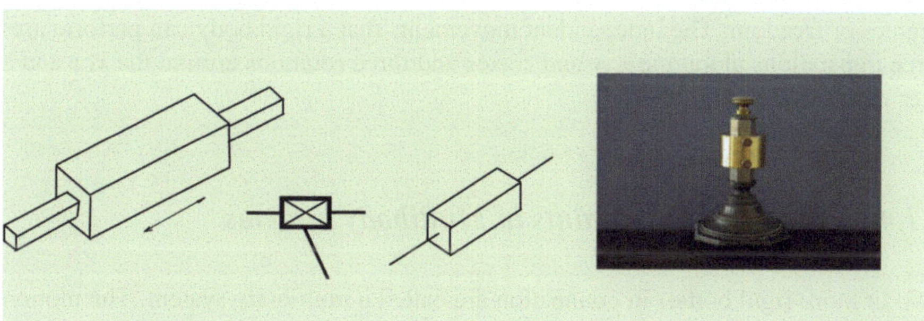

Fig. 5.17 Prismatic linkage (P)

limited, which is why at the moment most universities have given up such a discipline, encompassing the subjects covered in Mechanics, Machine Mechanics or Mechanics of Multibody Systems.

5.1.3 The Degrees of Freedom of a Rigid Body

We recapitulate some results presented on the kinematics of the rigid body due to the importance of these cases in practice.

5.1.3.1 The Number of Degrees of Freedom of a Rigid Body with Parallel Plane Motion

The number of degrees of freedom (DOF) of a rigid body is defined as the number of independent parameters needed to describe its motion.

When studying the kinematics of the rigid body, we saw that the number of degrees of freedom of a rigid body is 6. In the case of parallel plane motion, we saw that their number can be reduced to 3. The number of degrees of freedom can also be defined as the number of independent motion that a body can execute them. Thus, in the plane, a bar (or any body) can perform a translation along the x axis, a translation along the y axis and a rotation (Fig. 5.9).

5.1.3.2 The Number of Degrees of Freedom of a Rigid Body in General Motion

This was clarified in Chap. 3 and we saw that a rigid body in general motion has 6 degrees of freedom. The independent movements that a rigid body can perform are: three translations along the x, y and z axes and three rotations around the x, y and z axes respectively (Fig. 5.10).

5.1.4 Kinematic Constraints in Multibody Systems

Two or more rigid bodies in connection are called a multibody system. The motion of these bodies is limited by kinematic constraints. Kinematic constraints are links between the different rigid bodies and have the effect of decreasing the number of degrees of freedom of the system. The term linkage refers to the kinematic constraints between rigid bodies. Kinematic couples are divided into lower linkage and upper linkage depending on the way in which the two bodies are in contact.

5.2 Multibody Systems

5.2.1 Plane Mechanisms

There are two types of lower linkage in the theory of planar mechanisms: planar joints and pulleys [9–17]. A rigid body with parallel plane motion has three degrees of freedom, given by the possibility of translating in two directions and a rotation around an axis perpendicular to the plane of motion. If a cylindrical joint is introduced, the possibility of performing the two translations is suppressed, so two degrees of freedom are eliminated. If a translation coupler (slider) is introduced, a translation and rotation is suppressed, so again two degrees of freedom are eliminated. In both situations, the body will have only one degree of freedom. In the first case the body can perform only one rotation, in the second case only one translation (Figs. 5.11 and 5.12).

5.2.2 Lower Linkage in Spatial Mechanisms

There are six lower couples in the case of space mechanisms. These are:

spherical joint (S), plane linkage (E), cylindrical joint (C), rotation linkage/hinge (R), slider (C) and screw (H).

A spherical joint (Fig. 5.13) consists of two spheres with the same center. Two bodies connected by such a coupling can only perform rotational movements around the three axes but cannot perform any kind of translation. Neglecting the backlashes that may occur, such a coupling suppresses three degrees of freedom of the rigid. The number of degrees of freedom allowed becomes 3 in this case.

A flat pairs (Fig. 5.14) is constituted by two flat surfaces of some rigid, in contact. If the two bodies are connected in this way, one of them can have a translational movement in two directions and a rotation along an axis perpendicular to the surfaces in contact. Translation along a direction perpendicular to the surfaces and rotations around two axes located in the contact plane are prevented. It follows that the number of degrees of freedom left to a body following the connection through a planar linkage is 3.

A cylindrical linkage (Fig. 5.15) consists of two cylinders with the same axis. A cylindrical linkage allows only two movements: a rotation around the axis and a translation along the axis. The number of degrees of freedom allowed by such a couple is 2.

A revolute linkage (Fig. 5.16) consists of two cylindrical surfaces, which rotate with respect to each other. Translational movement along the axis is prevented. Such a coupling allows only one movement, the rotation around the axis. It is said to have one degree of freedom.

A prismatic linkage (Fig. 5.17) is made up of two bodies having a common axis along which the movement occurs. There is no relative rotation between the two

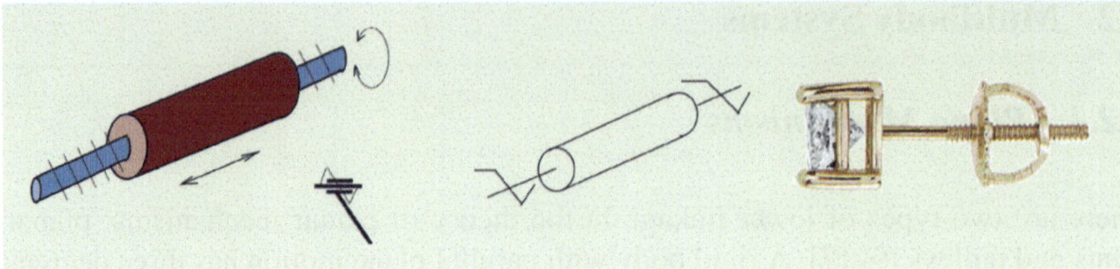

Fig. 5.18 Screw linkage (H)

bodies. There is only one translation along the common axis, so the system has only one degree of freedom.

A screw linkage (Fig. 5.18) is made up of two bodies, one of them having a screw movement with respect to the other. The translational and rotational movements are linked by a linear relationship and the system will in this case have only one degree of freedom.

5.2.3 Mobility

The mobility of a mechanism is represented by the number of relative movements that can exist between the constituent elements of the mechanism.

A plane body has three degrees of freedom. A link subtracts from the number of degrees of freedom of a body, as seen previously. If there are n moving bodies, l lower link (allowing one degree of freedom) and h upper link (allowing two degrees of freedom), the number of remaining degrees of freedom (or the mobility of the mechanism) is:

$$F = 3n - 2l - h, \tag{5.26}$$

where $F =$ total number of degrees of freedom in the mechanism.

Example 1 Consider the mechanism in Fig. 5.19. We have $n = 3$, $l = 4$ (A, B, C, D), $h = 0$. We calculate the number of degrees of freedom:

$$F = 3n - 2l - h = 3x3 - 2x4 - 1x0 = 1. \tag{5.27}$$

Note: D and E have the same function, so they are considered only once.

Example 2 Calculate the number of degrees of freedom of the mechanism in Fig. 5.20a. Figure 5.20b represents an application of the mechanism.

$$\mathrm{n} = 4, \quad \mathrm{l} = 4(\mathrm{A}, \mathrm{B}, \mathrm{C}, \mathrm{D}), \quad \mathrm{h} = 0,$$

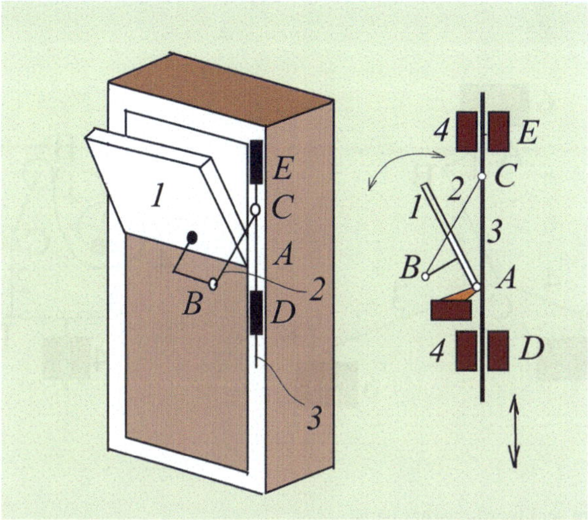

Fig. 5.19 A real mechanism

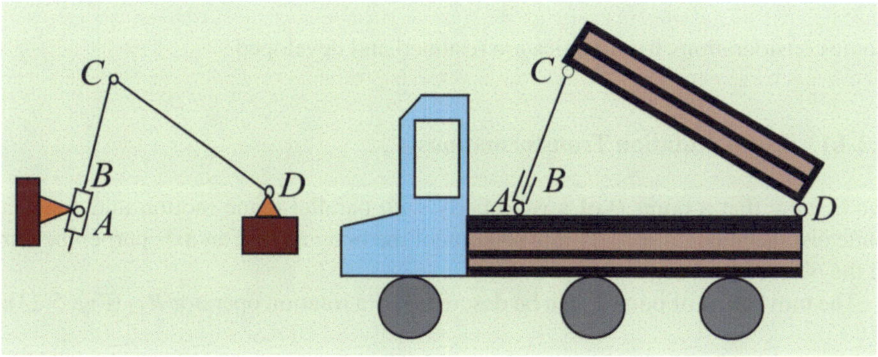

Fig. 5.20 The tipping mechanism of the truck

$$F = 3x3 - 2x4 - 1x0 = 1. \tag{5.28}$$

Example 3 Calculate the mobility of the mechanisms in Fig. 5.21.

For the mechanism in Fig. 5.21a: $n = 6, l = 7, h = 0$

$$F = 3x5 - 2x7 - 1x0 = 1. \tag{5.29}$$

For the mechanism in Fig. 5.21b: $n = 4, l = 3, h = 2$

$$F = 3x3 - 2x3 - 1x2 = 1. \tag{5.30}$$

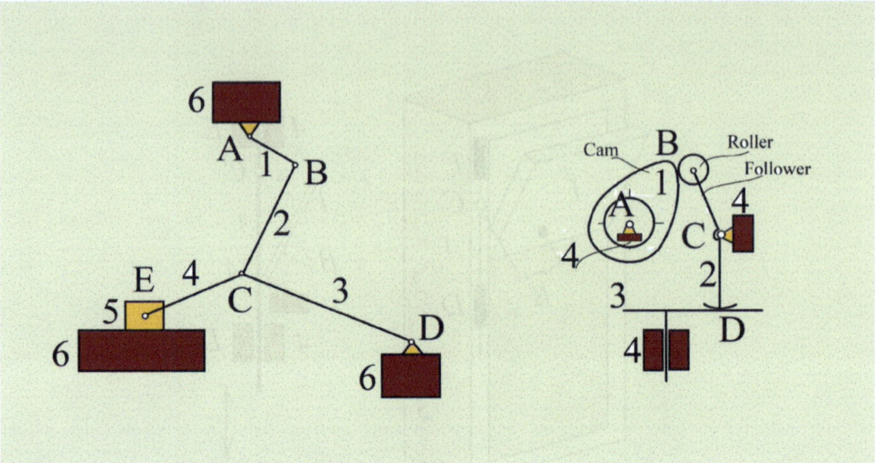

Fig. 5.21 Different real word mechanisms

5.2.4 *Finite Transformations*

Some considerations from Statics are resumed and developed.

5.2.4.1 Finite Rotation Transformations

We assume that a point P of a rigid body with parallel plane motion undergoes a finite displacement, due to a finite rotation of the body around an axis perpendicular to the plane, by an angle θ (Fig. 5.22).

The movement of point P can be described by a rotation operator $\boldsymbol{R}_{12}$ (Fig. 5.23):

$$\begin{Bmatrix} X_2 \\ Y_2 \\ 1 \end{Bmatrix} = [R_{12}] \begin{Bmatrix} X_1 \\ Y_1 \\ 1 \end{Bmatrix}; \tag{5.31}$$

Fig. 5.22 A point on a rigid body that has undergone a planar rotation

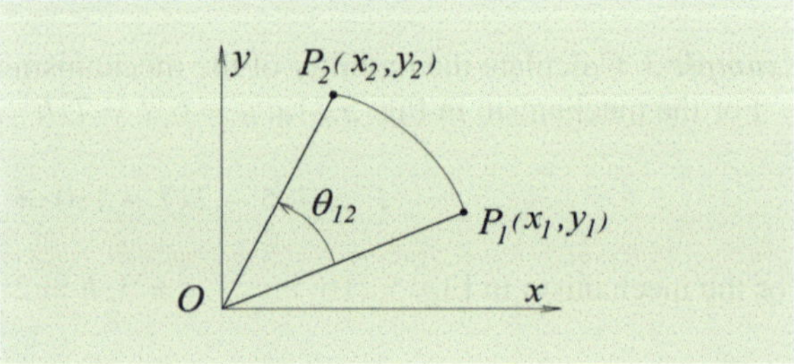

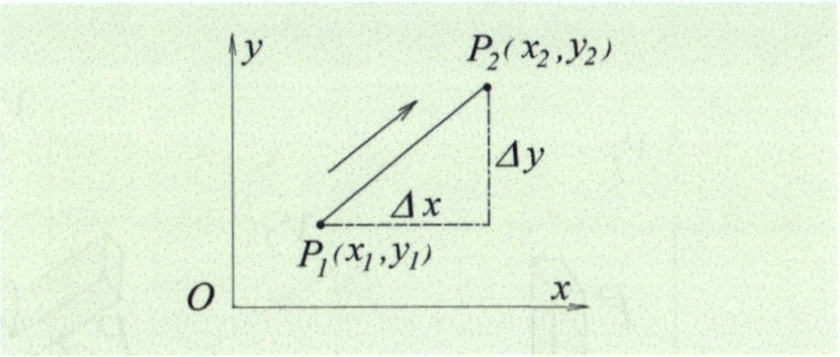

Fig. 5.23 Plane translation

where:

$$[R_{12}] = \begin{bmatrix} \cos\theta & -\sin\theta & 0 \\ \sin\theta & \cos\theta & 0 \\ 0 & 0 & 1 \end{bmatrix}. \tag{5.32}$$

5.2.4.2 Translation Transformations

A point P of a rigid performs a planar translation passing from P_1 to P_2 changing its coordinates with (Δx, Δy). The finite motion can be described by a translation operator $[T_{12}]$:

$$\begin{Bmatrix} X_2 \\ Y_2 \\ 1 \end{Bmatrix} = [T_{12}] \begin{Bmatrix} X_1 \\ Y_1 \\ 1 \end{Bmatrix}, \tag{5.33}$$

where:

$$[T_{12}] = \begin{bmatrix} 1 & 0 & \Delta x \\ 0 & 1 & \Delta y \\ 0 & 0 & 1 \end{bmatrix}. \tag{5.34}$$

5.2.4.3 Composition of Finite Displacements in the Plane

Let us consider a point P of the rigid body that changes its position following a rotation passing from P_1 to P_2' in a circular motion, after which it performs a translational movement passing from P_2' to P_2. We can represent these two steps by (Fig. 5.24):

$$\begin{Bmatrix} X_2' \\ Y_2' \\ 1 \end{Bmatrix} = [R_{12}] \begin{Bmatrix} X_1 \\ Y_1 \\ 1 \end{Bmatrix}; \tag{5.35}$$

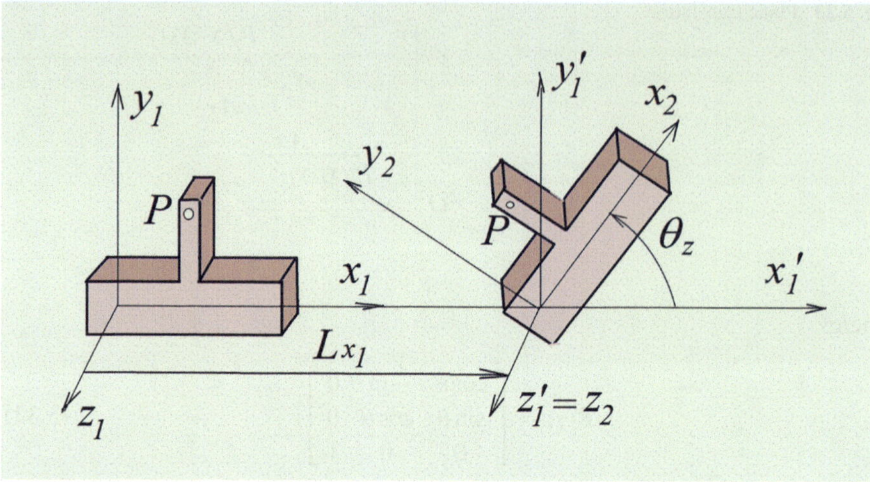

Fig. 5.24 Composition of finite displacements in the plane

and:

$$\begin{Bmatrix} X_2 \\ Y_2 \\ 1 \end{Bmatrix} = [T_{12}] \begin{Bmatrix} X_2' \\ Y_2' \\ 1 \end{Bmatrix}. \tag{5.36}$$

We compose these movements, obtaining:

$$\begin{Bmatrix} X_2 \\ Y_2 \\ 1 \end{Bmatrix} = [D_{12}] \begin{Bmatrix} X_1 \\ Y_1 \\ 1 \end{Bmatrix}; \tag{5.37}$$

where $[D_{12}]$ is the total displacement operator:

$$[D_{12}] = [T_{12}][D_{12}] = \begin{bmatrix} \cos\theta & -\sin\theta & L_{x1} \\ \sin\theta & \cos\theta & 0 \\ 0 & 0 & 1 \end{bmatrix}. \tag{5.38}$$

In the general case, when there is a translation with components along both axes, the total displacement operator will be:

$$[D_{12}] = [T_{12}][D_{12}] = \begin{bmatrix} \cos\theta_{12} & -\sin\theta_{12} & \Delta x \\ \sin\theta_{12} & \cos\theta_{12} & \Delta y \\ 0 & 0 & 1 \end{bmatrix}. \tag{5.39}$$

5.2.5 Transformations in Space

5.2.5.1 Space Rotation

Consider a rotation of a point around an axis $\boldsymbol{u}$ that passes through the origin of the coordinate system. The angle of rotation around this axis is θ. In this case the rotation operator is:

$$R_{\theta,u} = \begin{bmatrix} u_x^2 v_\theta + c_\theta & u_x u_y v_\theta - u_z s_\theta & u_x u_z v_\theta + u_y s_\theta & 0 \\ u_x u_y v_\theta + u_z s_\theta & u_y^2 v_\theta + c_\theta & u_y u_z v_\theta - u_x s_\theta & 0 \\ u_x u_z v_\theta - u_y s_\theta & u_y u_z v_\theta + u_x s_\theta & u_z^2 v_\theta + c_\theta & 0 \\ 0 & 0 & 0 & 1 \end{bmatrix}, \tag{5.40}$$

where u_x, u_y, u_z are the orthogonal projections of the $\boldsymbol{u}$ unitary axis on the x, y and z axes, respectively. It was noted:

$$s_\theta = \sin\theta\ ; \quad c_\theta = \cos\theta\ ; \quad v_\theta = 1 - \cos\theta. \tag{5.41}$$

5.2.5.2 Space Translation

If a point P of the rigid performs a translation, passing from P_1 to P_2 with the change of coordinates (Δx, Δy, Δz), the translation operator will describe this transformation:

$$[T] = \begin{bmatrix} 1 & 0 & 0 & \Delta x \\ 0 & 1 & 0 & \Delta y \\ 0 & 0 & 1 & \Delta z \\ 0 & 0 & 0 & 1 \end{bmatrix}. \tag{5.42}$$

5.2.5.3 Rotational and Translational Spatial Transformation

As in the plan case, by combining the two transformations we get:

$$R_{\theta,u} = \begin{bmatrix} u_x^2 v_\theta + c_\theta & u_x u_y v_\theta - u_z s_\theta & u_x u_z v_\theta + u_y s_\theta & Du_x \\ u_x u_y v_\theta + u_z s_\theta & u_y^2 v_\theta + c_\theta & u_y u_z v_\theta - u_x s_\theta & Du_y \\ u_x u_z v_\theta - u_y s_\theta & u_y u_z v_\theta + u_x s_\theta & u_z^2 v_\theta + c_\theta & Du_z \\ 0 & 0 & 0 & 1 \end{bmatrix}. \tag{5.43}$$

5.2.6 Denavit-Hartenberg's Notations

Denavit-Hartenberg notations are frequently used for the transformation of coordinate systems in the case of mechanisms and robots. They can be used to represent the transformation matrix between bodies as in Fig. 5.22 (Fig. 5.25).

In this figure:

z_{i-1} and z_i are the axes of two rotation joints;

θ_i represents the angle between axes x_{i-1} and x_i;

d_i is the distance between the origin of the coordinate system $x_{i-1}y_{i-1}z_{i-1}$ and the foot of the common perpendicular on the two axes;

a_i is the size of the common perpendicular;

α_i is the angle between the z_{i-1} and z_i axes;

The transformation matrix will be:

$$T_{(i-1)i} = \begin{bmatrix} \cos\theta_i & -\sin\theta_i\cos\alpha_i & \sin\theta_i\cos\alpha_i & a_i\cos\theta_i \\ \sin\theta_i & \cos\theta_i\cos\alpha_i & -\cos\theta_i\cos\alpha_i & a_i\sin\theta_i \\ 0 & \sin\alpha_i & \cos\alpha_i & d_i \\ 0 & 0 & 0 & 1 \end{bmatrix}. \quad (5.44)$$

This transformation matrix is also denoted by $T(a_i, \alpha_i, \theta_i, d_i)$.

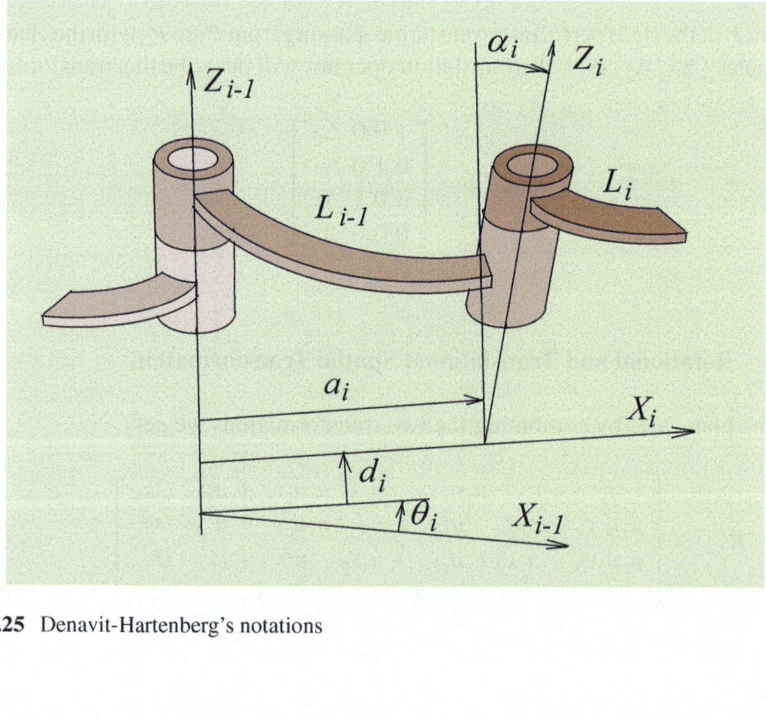

Fig. 5.25 Denavit-Hartenberg's notations

5.2.7 *Application of Matrix Transformations to Kinematic Chains*

A kinematic chain is composed of linked rigid bodies. The matrix method can be used to obtain the equations of motion. If the elements of the mechanical system are linked to form a closed loop, the composition of all transformation matrices will result in the unity matrix. If the system consists of n bodies we have:

$$T_{12}T_{23}\ldots T_{(n-1)n} = E. \tag{5.45}$$

The presented formula allows the development of a method for solving the mechanisms, namely the method of loops equations.

5.3 Method of Loops Equations

We present the method of independent loops following the course [4]. A more complete presentation is given in the book [5]. The method refers to solving mechanisms, defined as closed kinematic chains. A mechanism is defined by most authors as consisting of at least one closed kinematic chain. (Although this definition significantly reduces the field of mechanical systems that can be studied. As a result, the notion of a multibody mechanical system was introduced, which allows the study of both closed and open kinematic chains). If several kinematic chains can be identified within a mechanism, among them there is a minimum number that determines the mechanical problem. The idea of the method is to use the kineto-static analogy for an entire kinematic chain.

Thus, the determination of the angular velocity and the velocity of a point belonging to the last body in a series of "n" bodies that move relative to each other, comes down to the reduction at that point of the system of relative angular velocities—considered as sliding vectors, and of relative linear velocities system-considered as free vectors. The relationships will be obtained.

$$\vec{\omega}_{n0} = \sum_{i=1}^{n} \vec{\omega}_{i,i-1}; \tag{5.46}$$

$$\vec{v}_{n0}^{A} = \sum_{i=1}^{n} \vec{v}_{i,i-1} + \sum_{i=1}^{n} \vec{\omega}_{i,i-1}\, x\, \overrightarrow{O_iA}\ . \tag{5.47}$$

In the following, we will consider a closed kinematic chain made up of a number of "n" bodies connected to each other, which we will number ascending, in the order in which they follow in the kinematic chain 1, 2, i, ..., n. The connections between the bodies are made through the kinematic links A, B, ..., M. The relative angular velocities and the relative linear velocities were represented in the kinematic links. The absolute speed of point P belonging to body i, expressed as a function of the absolute speed of body A belonging to body i is (Fig. 5.26):

$$\vec{v}_{i,0}^{P} = \vec{v}_{i,0}^{A} + \vec{v}_{Pi,Ai}, \tag{5.48}$$

where we also have the relation:

$$\vec{v}_{i,0}^{A} = \vec{v}_{j,0}^{A} + \vec{v}_{Ai,Aj}. \tag{5.49}$$

With $\vec{v}_{Pi,Ai}$ is denoted the relative speed of point P to A, both belonging to body i, i.e.:

$$\vec{v}_{Pi,Ai} = \vec{\omega}_{i0}\, x\, A_iP_i; \tag{5.50}$$

where:

$$\vec{\omega}_{i,0} = \vec{\omega}_{j,0} + \vec{\omega}_{i,j}. \tag{5.51}$$

With $\vec{v}_{Ai,Aj}$ is the relative speed of point A on body I compared to point A on body j.

If the previous relations are taken into account, the expression of the absolute speed of the point P on the body i will become:

$$\vec{v}_{i,0}^{P} = \vec{v}_{j,0}^{A} + \vec{v}_{AiAj} + \vec{\omega}_{i0}\, x\, \overrightarrow{A_iP_i}. \tag{5.52}$$

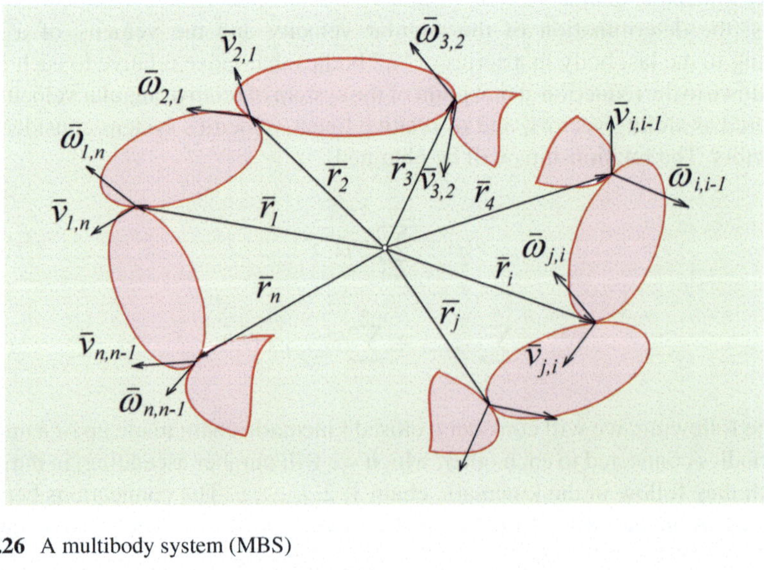

Fig. 5.26 A multibody system (MBS)

The position vector $\overrightarrow{A_iP_i} = \overrightarrow{AP}$ can be expressed as a function of the position vectors of points A and P relative to the fixed frame O:

$$\overrightarrow{A_iP_i} = \vec{r}_P - \vec{r}_A. \tag{5.53}$$

It results:

$$\vec{\omega}_{i,0} x \overrightarrow{A_iP_i} = \vec{\omega}_{i,0}\, x\, (\vec{r}_P - \vec{r}_A). \tag{5.54}$$

Summing up for all the bodies in a closed kinematic chain the relations of the type (5.54) it obtains:

$$\begin{aligned}&\vec{\omega}_{i,0}\, x\, (\vec{r}_P - \vec{r}_A) + \vec{\omega}_{j,0}\, x\, (\vec{r}_A - \vec{r}_B) + \vec{\omega}_{k,0} x\, (\vec{r}_B - \vec{r}_C)\\ &\quad + \cdots + \vec{\omega}_{n0}\, x\, (\vec{r}_Q - \vec{r}_P) = (\vec{\omega}_{j,0} - \vec{\omega}_{i,0})\, x\, \vec{r}_A + (\vec{\omega}_{k,0} - \vec{\omega}_{j,0})\, x\, \vec{r}_B\\ &\quad + \cdots + (\vec{\omega}_{i,0} - \vec{\omega}_{n,0})\, x\, \vec{r}_P.\end{aligned} \tag{5.55}$$

Because:

$$\vec{\omega}_{i0} - \vec{\omega}_{i0} = \vec{\omega}_{ij}; \tag{5.56}$$

which represents the relative angular velocity, and reversing the order in the vector product, the sum from relation (5.55) can also be written:

$$\vec{r}_A\, x\, \vec{\omega}_{ij}. \tag{5.57}$$

Writing the relations (5.51) and (5.52) for all the bodies of a closed kinematic chain and taking into account the results (5.54)–(5.57), the following relations are obtained:

$$\sum \vec{\omega}_{ij} = 0; \tag{5.58}$$

$$\sum \vec{v}_{AiAj} + \sum \vec{r}_A\, x\, \vec{\omega}_{ij} = 0. \tag{5.59}$$

The vector relations (5.58) and (5.59) are used to determine the velocities and are analogous to the equilibrium equations (for forces and moments) written for a system of solidified bodies. Thus, the static-kinematic analogy is highlighted.

In the case of problems that are solved with this method, the vector relations (5.58) and (5.59) are projected onto the axes of a fixed axis system and equations are obtained that are analogous to the equilibrium equations in statics. In this case, the system load consists of $\vec{\omega}_{ij}$ and $\vec{v}_{AiAj}$.

For the analysis of the accelerations, the relations obtained in the study of the relative motion of rigid bodies are used:

$$\vec{\varepsilon}_{n0} = \sum_{i=1}^{n} \vec{\varepsilon}_{i,i-1} + \sum_{i=1}^{n} \sum_{j=1}^{i-1} \left(\vec{\omega}j_{j,j-1} \times \vec{\omega}_{i,i-1}\right); \tag{5.60}$$

$$\begin{aligned} \vec{a}_{n,0}^{M} = & \sum_{i=1}^{n} \vec{a}_{i,i-1} + \sum_{i=1}^{n} \vec{\varepsilon}_{i,i-1} \times \vec{r}_i + \sum_{i=1}^{n} \left[\vec{\omega}_{i,i-1} \times \left(\vec{\omega}_{i,i-1} \times \vec{r}_i\right)\right] \\ & + 2 \sum_{i=1}^{n} \sum_{j=1}^{i-1} \vec{\omega}_{i-1,0} \times \left(\vec{v}_{i,i-1} + \vec{\omega}_{j,j-1} \times \vec{r}_i\right). \end{aligned} \tag{5.61}$$

If the same path is followed as for determining the velocities, the absolute acceleration of point P belonging to body i depending on the acceleration of point A on the same body is:

$$\vec{a}_{i0}^{P} = \vec{a}_t + \vec{a}_c + \vec{a}_r = \vec{a}_{j0}^{A} + \vec{a}_{AiAj}^{t} + \vec{a}_{AiAj}^{n} + \vec{a}_{AiAj}^{c} + \vec{a}_{PiAi}; \tag{5.62}$$

where the relative acceleration of point P to A is calculated with:

$$\vec{a}_{PiAi} = \vec{\varepsilon}_{i0}\, x\, \overrightarrow{A_i P}_i + \vec{\omega}_{i0}\, x\, (\vec{\omega}_{i0}\, x\, \overrightarrow{A_i P}_i). \tag{5.63}$$

In expression (5.63) we apply (5.60) in the form:

$$\vec{\varepsilon}_{i0} = \vec{\varepsilon}_{j0} + \vec{\varepsilon}_{ij} + \vec{\omega}_{j0}\, x\, \vec{\omega}_{ij}. \tag{5.64}$$

If we write (5.63) and (5.64) and take into account (5.54), we obtain the general relations that can be used for the study of accelerations:

$$\sum \vec{\varepsilon}_{ij} + \sum \vec{\omega}_{j0}\, x\, \vec{\omega}_{ij} = 0; \tag{5.65}$$

$$\begin{aligned} & \sum \vec{a}_{AiAj}^{t} + \sum \vec{r}_A\, x\, \left(\vec{\varepsilon}_{ij} + \vec{\omega}_{j0}\, x\, \vec{\omega}_{ij}\right) + \sum a_{AiAj}^{n} + \sum \vec{a}_{AiAj}^{c} \\ & + \sum \vec{\omega}_{i0}\, x \left(\vec{\omega}_{i0}\, x\, \overrightarrow{AP}\right) = 0. \end{aligned} \tag{5.66}$$

We mention that in the case of planar mechanisms the $\vec{\omega}_{ij}$ and $\vec{\omega}_{i0}$ vectors are parallel, so their cross product is zero. In practical applications, the vector Eqs. (5.65) and (5.66) are projected onto a conveniently chosen fixed axis system and the scalar equilibrium equations are obtained.

If a mechanism has several independent loops, the method is applied for each loop separately.

5.4 The Method of Loop-Closure Equations

A contour made up of segments, connected to each other, can be attached to a set of bodies. Each segment is attached to a specific rigid body and participates in movement alongside it. If the vector relations that define a vector contour are derived, relations between the kinematic elements of the different bodies are obtained. The method for planar mechanisms is presented.

The algorithm that is based on this method was developed using graph theory. This theory was used in the analysis of mechanisms in two different directions:

- A direction in which the moving elements of the system are considered nodes of the graph and the links branches. In this approach, an equivalence is made with Kirkoff's equations for electrical circuits, the dynamic equilibrium equations for the elements that make up the mechanical system being similar to the balance of currents in nodes, and the kinematic conditions being corresponding to the equations written for a loop. This type of analysis led to the method of independent cycles.
- Another approach would be to consider the links between elements as nodes and the vector defined by two links as a branch of the graph. This type of description, geometric, has been addressed less in the literature. In what follows, this type of topological description of a mechanical system with links is followed. The plane case, presented in the course, can be extended to spatial mechanisms.

5.4.1 Topological Description of the Mechanism

Equivalent Kinematic Mechanism

In what follows, a graph will be attached to each mechanism. In order to do this operation, the first step is to build a mechanism made up of bars only, equivalent from a mechanical point of view to the given mechanism.

Let us consider the plane mechanism represented in Fig. 5.27. The connection between two elements is made by joints or slider.

We construct a kinematic mechanism equivalent to the one given in the following way:

I. We replace any plate-type mobile element (Fig. 5.28) with the rigid structure, made up of triangles. In this way, "false" fundamental loops (independent cycles) are introduced into the equations describing the kinematic state of the system, but which facilitate the representation of the equations.
II. Connect the points that make the connection with the fixed element through bar-type elements that close a polygon (Fig. 5.29) and then eliminate one of these elements (it results from the vector sum with changed sign of all the others). In this way, "false" commender elements are introduced that can change their length, but not their position, so always the angular velocity and angular acceleration of such an element are zero.

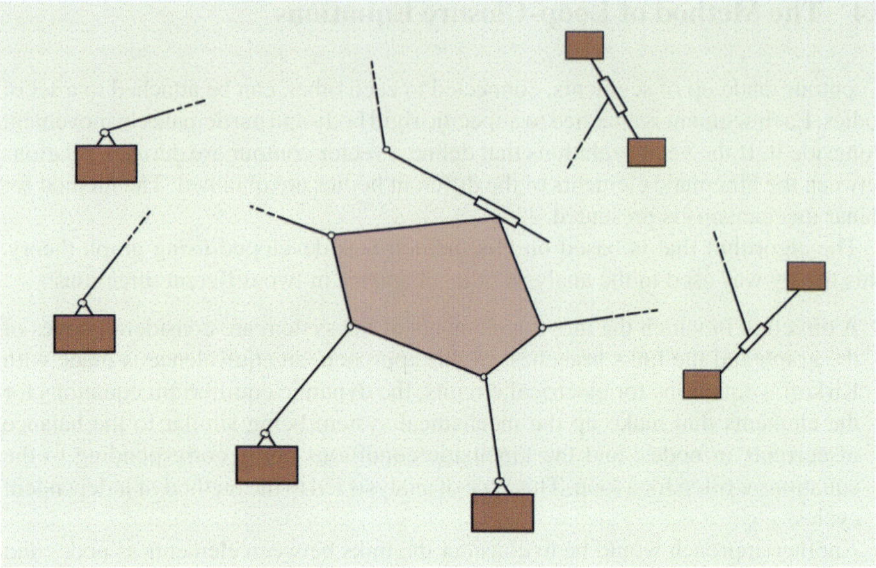

Fig. 5.27 A general mechanism

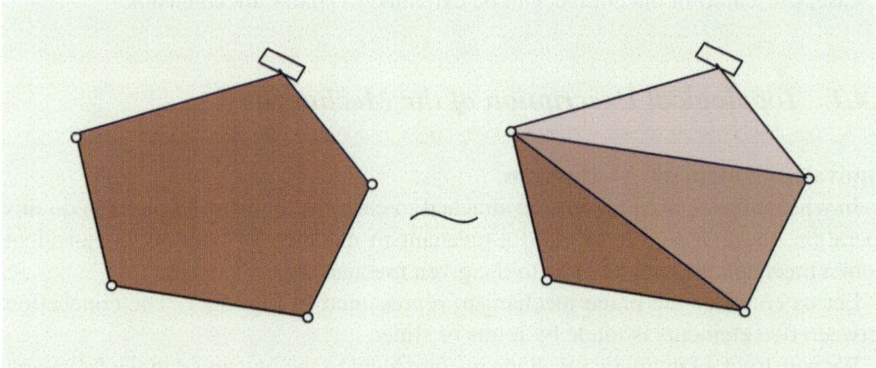

Fig. 5.28 Decomposing a plate into triangles

Through these two constructions, a kinematically equivalent mechanism to the given one is obtained, but made up only of bars (Fig. 5.29). At the time when we propose to determine the kinematic state of the mechanism elements, we consider known all the lengths of the constitutive bars of the equivalent kinematic mechanism and the angles made by them with the Ox axis of a chosen xOy system.

By attaching to each bar element a vector with the end in one link and the tip in the other, we build a graph that has as nodes the links of the mechanism, as sides the bars of the mechanism and the meaning of each side is given by the direction of the vector attached to the corresponding bar (Fig. 5.30).

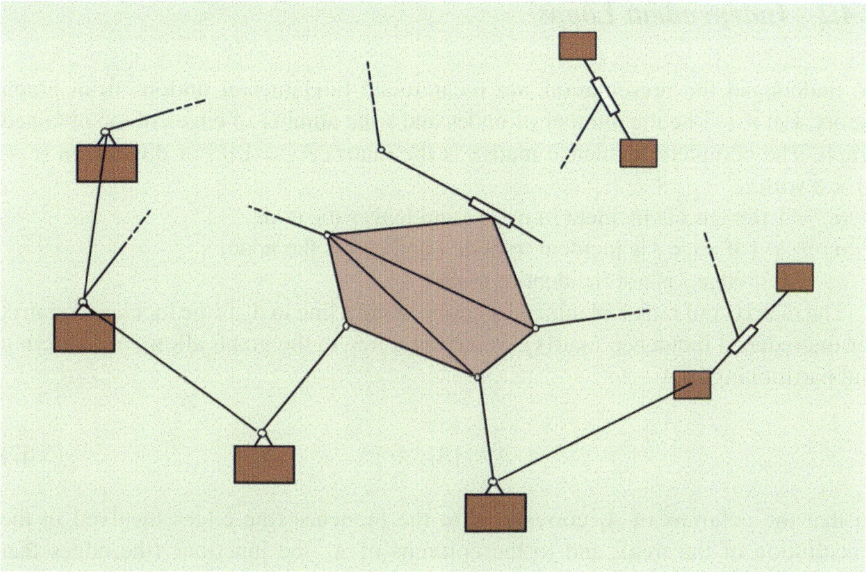

Fig. 5.29 Kinematic equivalent mechanism

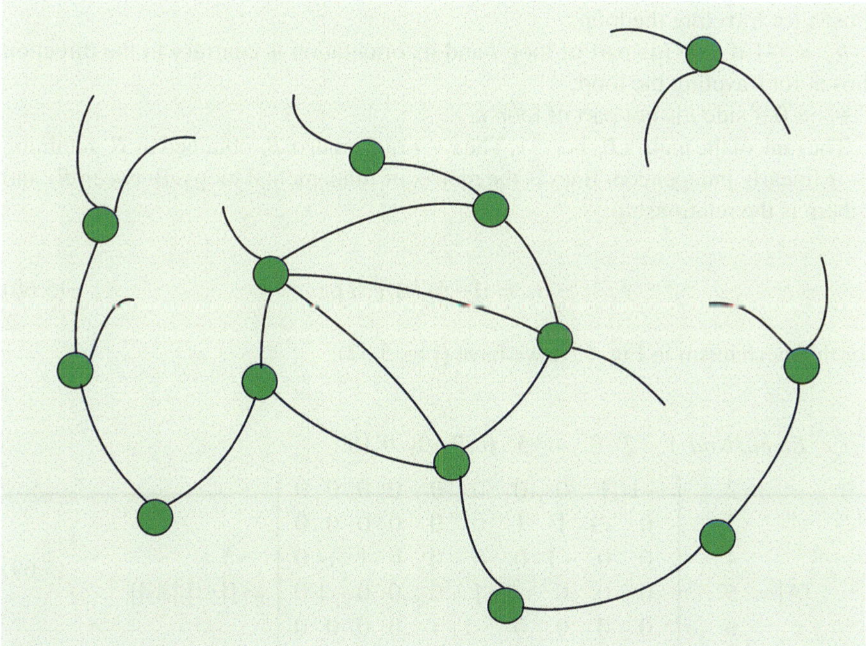

Fig. 5.30 The attached graph

5.4.2 Independent Loops

To understand the presentation, we recapitulate fundamental notions from graph theory. Let $r+1$ be the number of nodes and s the number of edges of the obtained graph. The complete incidence matrix is the matrix $A_a = [a_{ij}]$ of dimension $(r+1) \times s$ with:

$a_{ij} = 1$ if edge j is incident to node i and leaves the node;

$a_{ij} = -1$ if edge j is incident to node i and enters the node;

$a_{ij} = 0$ if edge j is not incident to node i.

The matrix A of rank r obtained by suppressing a line in A_a is the incidence matrix (or the reduced incidence matrix). Selecting a tree in the graph allows a reordering and partitioning of A:

$$A = [A_t \vdots A_l]; \tag{5.67}$$

so that the columns of A_t correspond to the branches (the edges involved in the constitution of the tree), and to the columns of A_l the junctions (the edges that attached to the chosen tree give the initial graph).

The complete loop matrix is the matrix: $B_a = [b_{ij}]$ with:

$b_{ij} = 1$ if side j is part of loop i and its orientation coincides with the direction chosen for traveling the loop;

$b_{ij} = -1$ if side j is part of loop i and its orientation is contrary to the direction chosen for traveling the loop;

$b_{ij} = 0$ if side j is not part of loop i.

The rank of the matrix B_a is $s-r$. The $s-r$ rank matrix B_f obtained by B_a retaining $s-r$ linearly independent lines is the matrix of fundamental loops. Between B_f and A there is the relationship:

$$B_f = [[-A_t^{-1} A_l]^T \vdots I]. \tag{5.68}$$

For the mechanism in Fig. 5.31 we have (Fig. 5.32):

$$\begin{array}{c} \textit{Elem./Nod} \;\; 1 \quad 2 \quad 3 \quad 4 \quad 5 \quad 6 \quad 7 \quad 8 \quad 9 \quad 10 \\ [A] = \begin{array}{c} 2 \\ 3 \\ 4 \\ 5 \\ 6 \\ 7 \\ 8 \end{array} \begin{bmatrix} -1 & 1 & 0 & 0 & 0 & 0 & 0 & 0 & 0 & 0 \\ 0 & -1 & 1 & 1 & 0 & 0 & 0 & 0 & 0 & 0 \\ 0 & 0 & -1 & 0 & 0 & 0 & 0 & 1 & 1 & 0 \\ 0 & 0 & 0 & -1 & 1 & 0 & 0 & 0 & -1 & 0 \\ 0 & 0 & 0 & 0 & -1 & 1 & 0 & 0 & 0 & 0 \\ 0 & 0 & 0 & 0 & 0 & -1 & 1 & 0 & 0 & 0 \\ 0 & 0 & 0 & 0 & 0 & 0 & -1 & 0 & 0 & 1 \end{bmatrix} = [[A_t]\,[A_l]] \end{array} \tag{5.69}$$

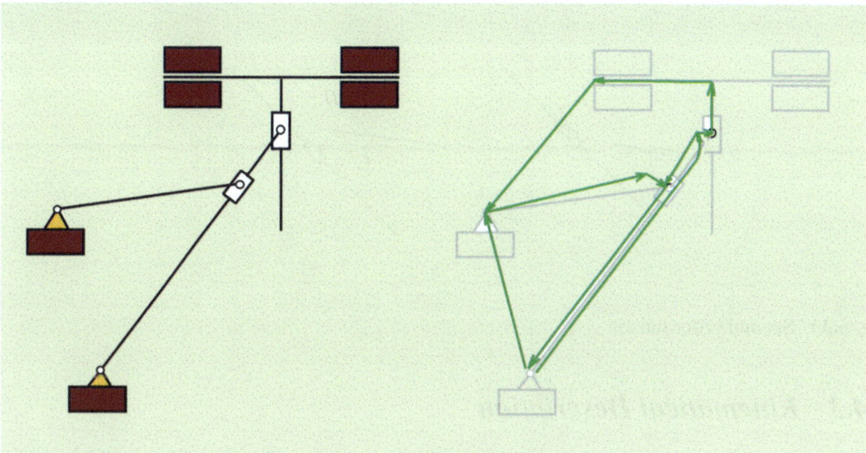

Fig. 5.31 Mechanism and kinematic equivalent mechanism

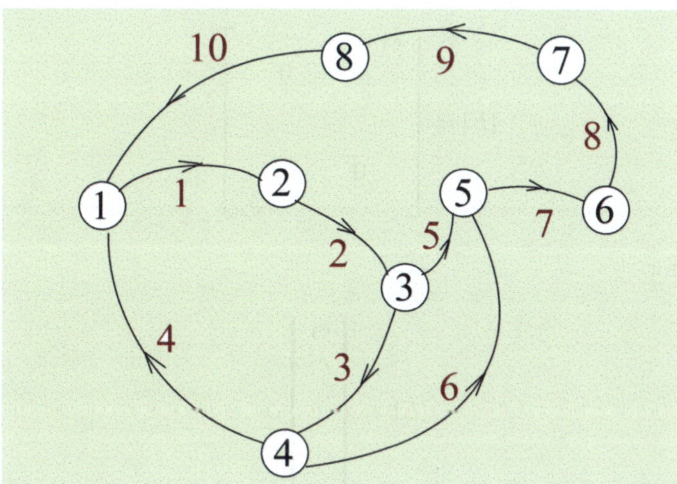

Fig. 5.32 Associated graph

After calculations, the matrix of fundamental loops results:

$$[B_f] = \begin{bmatrix} 1 & 1 & 1 & 0 & 0 & 0 & 0 & 1 & 0 & 0 \\ 0 & 0 & 1 & -1 & 0 & 0 & 0 & 0 & 1 & 0 \\ 1 & 1 & 0 & 1 & 1 & 1 & 1 & 0 & 0 & 1 \end{bmatrix}. \tag{5.70}$$

In this example, in order to be able to do the analysis, we made the substitution in Fig. 5.33.

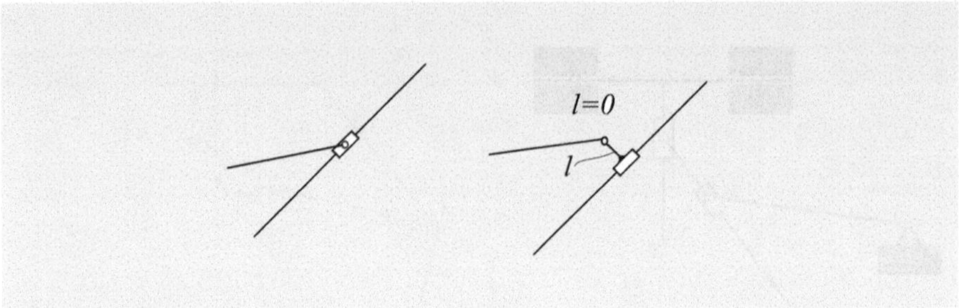

Fig. 5.33 Second order liaison

5.4.3 *Kinematical Description*

Notations

Let s be a set of sizes $(e_1, e_2, \ldots, e_s)$. With these we build the diagonal matrix:

$$[E] = \begin{bmatrix} e_1 & & & \\ & e_2 & 0 & \\ & & \ddots & \\ & 0 & & \ddots & \\ & & & & e_s \end{bmatrix}, \tag{5.71}$$

and the vector

$$\{E\} = \begin{Bmatrix} e_1 \\ e_2 \\ \vdots \\ \vdots \\ e_s \end{Bmatrix}. \tag{5.72}$$

Let $(l_1, l_2, \ldots, l_s)$ the lengths of the sides of the mechanism be In the above way we build the diagonal matrix $[L]$ and the vector $\{L\}$.

If $\theta_1, \theta_2, \ldots, \theta_s$ are the angles made by the vectors with the Ox axis, we construct with $(\cos\theta_1, \cos\theta_2, \ldots, \cos\theta_s)$ the matrix $[C]$ and the vector $\{C\}$, and with $(\sin\theta_1, \sin\theta_2, \ldots, \sin\theta_s)$ the matrix $[S]$ and the vector $\{S\}$. Similarly, if we denote by $\omega_i = \frac{d\theta_i}{dt}$, $v_i = \frac{dl_i}{dt}$, $\varepsilon_i = \frac{d\omega_i}{dt}$, $a_i = \frac{dv_i}{dt}$ we obtain the matrices $[\omega]$, $[v]$, $[\varepsilon]$, $[a]$ and the vectors $\{\omega\}$, $\{v\}$, $\{\varepsilon\}$, $\{a\}$. Consider:

$$[F] = \begin{bmatrix} f_1 & & & & \\ & f_2 & & 0 & \\ & & \ddots & & \\ & 0 & & \ddots & \\ & & & & f_s \end{bmatrix} \text{ and } \{F\} = \begin{Bmatrix} f_1 \\ f_2 \\ \vdots \\ \vdots \\ f_s \end{Bmatrix}. \tag{5.73}$$

It is immediately verified that:

$$[F][E] = [E][F]; \tag{5.74}$$

$$\{F\}^T[E] = \{E\}^T[F]. \tag{5.75}$$

Let e_i and f_i be time-varying. We make the notations:

$$[\dot{E}] = \frac{d[E]}{dt}; \ \{\dot{E}\} = \frac{d\{E\}}{dt}; \ [\dot{F}] = \frac{d[F]}{dt}; \ \{\dot{F}\} = \frac{d\{F\}}{dt}. \tag{5.76}$$

The following relation is checked:

$$\begin{aligned} \frac{d\left(\{F\}^T[E]\right)}{dt} = \frac{d\left(\{E\}^T[F]\right)}{dt} &= \{\dot{F}\}^T[E] + \{F\}^T[\dot{E}] \\ &= \{\dot{E}\}^T[F] + \{E\}^T[\dot{F}]. \end{aligned} \tag{5.77}$$

Loop-closure equations

We isolate from the mechanism the loop p composed of sides a, b, …, l (Fig. 5.30). The loop p corresponds to the line p in the matrix of fundamental loops which we will denote by B_p (Fig. 5.34).

$$\begin{aligned} &Elem. \, 1 \; 2 \ldots \; a \; \ldots \; b \; \ldots \; l \; \ldots \; s \\ &B_p = \left[0 \; 0 \ldots 0 \; a \; 0 \ldots 0 \; b \; 0 \ldots 0 \; l \; 0 \ldots s \right] \end{aligned} \tag{5.78}$$

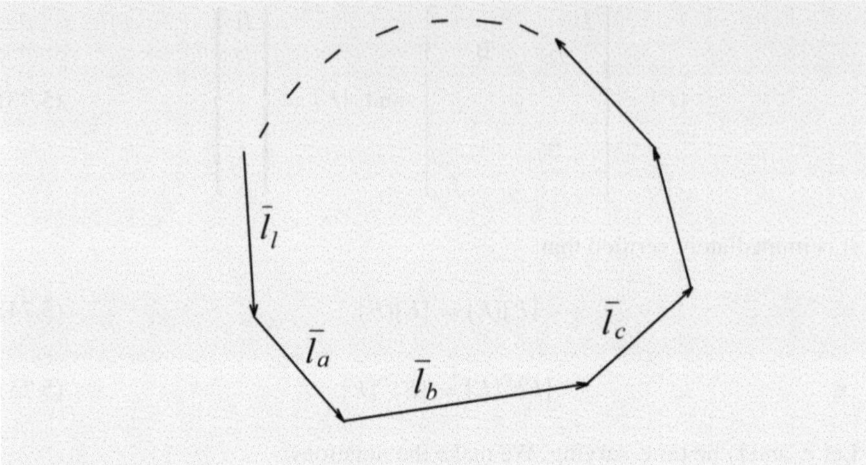

Fig. 5.34 A kinematic loop

The loop-closure equation will be written:

$$\vec{l}_a + \vec{l}_b + \cdots + \vec{l}_l = 0; \tag{5.79}$$

or, projecting along two axes:

$$\begin{aligned} l_a \cos\theta_a + l_b \cos\theta_b + \cdots \quad \cdots + l_l \cos\theta_l = 0 \\ l_a \sin\theta_a + l_b \sin\theta_b + \cdots \quad \cdots + l_l \sin\theta_l = 0. \end{aligned} \tag{5.80}$$

Let be $\{L_x\}$ the vector $\left\{ l_1 \cos\theta_1 \; l_2 \cos\theta_2 \; \ldots \; l_s \cos\theta_s \right\}^T$ and $\left\{L_y\right\}$ the vector $\left\{ l_1 \sin\theta_1 \; l_2 \sin\theta_2 \; \ldots \; l_s \sin\theta_s \right\}^T$. The contour equation for the p-loop can be written:

$$\left[B_p\right]\{L_x\} = \{0\}; \quad \left[B_p\right]\left\{L_y\right\} = \{0\}. \tag{5.81}$$

Doing $p = 1, 2, \ldots, r - s$ and because:

$$\left[B_f\right] = \begin{bmatrix} B_1 \\ B_2 \\ \vdots \\ B_{r-s} \end{bmatrix}, \tag{5.82}$$

the contour equations for the entire mechanism are written:

$$\left[B_f\right]\{L_x\} = \{0\}; \quad \left[B_f\right]\left\{L_y\right\} = \{0\}. \tag{5.83}$$

It is noted that:

$$\{L_x\} = [L]\{C\} = [C]\{L\}; \quad \{L_y\} = [L]\{S\} = [S]\{L\}\,. \tag{5.84}$$

Loop-closure equations for speeds and accelerations

Because:

$$[\dot{C}] = -[S][\omega]; \quad [\dot{S}] = [C][\omega]; \quad [\dot{L}] = [v], \tag{5.85}$$

we have:

$$\{\dot{L}_x\} = -[L][S]\{\omega\} + [C]\{v\}; \quad \{\dot{L}_y\} = [L][C]\{\omega\} + [S]\{v\}. \tag{5.86}$$

We can therefore write the loop-closure equations for the velocities obtained if the relations (5.86) are taken into account:

$$[B_f]\{\dot{L}_x\} = \{0\}; \quad [B_f]\{\dot{L}_y\} = \{0\}. \tag{5.87}$$

We then have:

$$\begin{cases} \{\ddot{L}\}_x = -[L][C][\omega]\{\omega\} - [L][S]\{\varepsilon\} - 2[S][v]\{\omega\} + [C]\{a\} \\ \{\ddot{L}\}_y = -[L][S][\omega]\{\omega\} + [L][C]\{\varepsilon\} - 2[C][v]\{\omega\} + [S]\{a\} \end{cases}. \tag{5.88}$$

Relations (5.88) allows to write the equations to determine the accelerations:

$$[B_f]\{\ddot{L}_x\} = \{0\}; \quad [B_f]\{\ddot{L}_y\} = \{0\}. \tag{5.89}$$

5.4.4 Kinematic Analysis of the Mechanism

The number of equations provided by (5.86) is 2x$(s-r)$, so 2x$(s-r)$ unknowns can be determined—velocities (angular and linear). It follows that in order for the motion of the mechanism to be determined, the other s – 2x$(s$ – 1) quantities, velocities appearing in relations (5.86) must be known.

The reasoning remains valid also for accelerations (Eq. (5.88)). If two elements i and j are connected to each other by a slide it results $\omega_i = \omega_j$. This allows reducing the number of angular velocities in the vector $\{\omega\}$. So it can be written:

$$\{\omega\} = [R_2]\{\omega^*\}; \tag{5.90}$$

where $\{\omega^*\}$ contains independent angular velocities.

If element i has two joints at the ends or one joint and a rigidly connected slide then:

$$\frac{dl_i}{dt} = 0. \tag{5.91}$$

There are therefore a number of null elements in the vector $\{v\}$. We denote $\{v^*\}$ the vector of nonzero components. We have:

$$\{v\} = [R_1]\{v^*\}. \tag{5.92}$$

Relations (5.90) and (5.92) remain valid for accelerations:

$$\begin{aligned}\{\varepsilon\} &= [R_1]\{\varepsilon^*\};\\ \{a\} &= [R_1]\{a^*\}.\end{aligned} \tag{5.93}$$

Relations (5.86) become:

$$\{\dot{L}_x\} = -[L][S][R_2]\{\omega^*\} + [C][R_1]\{v^*\} = \left[-[L][S][R_2]\ [C][R_1]\right]\begin{Bmatrix}\omega^*\\ v^*\end{Bmatrix} \tag{5.94}$$

$$\{\dot{L}_y\} = [L][C][R_2]\{\omega^*\} + [S][R_1]\{v^*\} = \left[[L][C][R_2]\ [S][R_1]\right]\begin{Bmatrix}\omega^*\\ v^*\end{Bmatrix} \tag{5.95}$$

so the condition equations for velocities (5.86) become:

$$\begin{bmatrix}-[B_f][L][S][R_2] & [B_f][C][R_1]\\ -[B_f][L][C][R_2] & [B_f][S][R_1]\end{bmatrix}\begin{Bmatrix}\omega^*\\ v^*\end{Bmatrix} = 0. \tag{5.96}$$

Rearrange the lines of the vector $\begin{Bmatrix}\omega^*\\ v^*\end{Bmatrix}$ so that the unknown quantities are on the first 2x$(s - r)$ positions and we note:

$$\{\dot{q}\} = [\dot{q}_1\ \dot{q}_2\ \ldots\ \dot{q}_{2x(s-r)}\ \dot{q}_{2x(s-r)+1}\ \ldots\ \dot{q}_1\ ]^T = \begin{Bmatrix}\{\dot{q}\}_n\\ \cdots\\ \{\dot{q}\}_c\end{Bmatrix}, \tag{5.97}$$

where $\dot{q}_i$ is an element of ω^* or v^*, $i = 1, 2, \ldots, 2x(s - r)$ indicates the unknown elements, $i = 2x(s - r) + 1, \ldots, s$ indexed the known elements. By reordering the columns:

$$\begin{bmatrix}-[B_f][L][S][R_2] & [B_f][C][R_1]\\ -[B_f][L][C][R_2] & [B_f][S][R_1]\end{bmatrix}, \tag{5.98}$$

corresponding to the reorderings through which $\left\{\begin{matrix}\omega^* \\ v^*\end{matrix}\right\}$ it passed in $\{\dot{q}\}$ and noting this matrix after reorderings with $[P]$, the relations (5.87) are written:

$$[P]\{\dot{q}\} = 0. \tag{5.99}$$

Partitioning $[P] = [P_n\ P_c]$ such that multiplication of partitioned matrices:

$$[P_n\ P_c]\left\{\begin{matrix}\{\dot{q}\}_n \\ \{\dot{q}\}_c\end{matrix}\right\} = 0; \tag{5.100}$$

to make sense, the condition equation for the velocities becomes:

$$[P]_n\{\dot{q}\}_n + [P]_c\{\dot{q}\}_c = 0. \tag{5.101}$$

From here the vector of the unknown velocities is determined:

$$\{\dot{q}\}_n = -[P]_n^{-1}[P]_c\{\dot{q}\}_c. \tag{5.102}$$

The vectors can now be easily constructed:

$$\{\omega\}^T[\omega] = [\omega_1^2\ \ \omega_2^2\ \ \ldots\ \ \omega_s^2]^T; \tag{5.103}$$

and:

$$\{v\}^T[\omega] = [v_1\omega_1\ \ v_2\omega_2^2\ \ \ldots\ \ v_s\omega_s^2]^T. \tag{5.104}$$

We denote:

$$\{q\}_2 = \{\omega\}^T[\omega]; \quad \{q\}_{cc} = \{v\}^T[\omega]. \tag{5.105}$$

With these notations, relations (5.88) become:

$$\begin{aligned}\{\ddot{L}_x\} &= -[L][C]\{q\}_2 - [L][S][R_2]\{\varepsilon^*\} - 2[S]\{q\}_{cc} + [C][R_1]\{a^*\}\\ \{\ddot{L}_y\} &= -[L][S]\{q\}_2 + [L][C][R_2]\{\varepsilon^*\} - 2[C]\{q\}_{cc} + [S][R_1]\{a^*\},\end{aligned} \tag{5.106}$$

and the condition conditions for accelerations (5.89) become:

$$\begin{aligned}&\begin{bmatrix}-[B_f][L][S][R_2] & [B_f][C][R_1]\\ -[B_f][L][C][R_2] & [B_f][S][R_1]\end{bmatrix}\left\{\begin{matrix}\varepsilon^* \\ a^*\end{matrix}\right\} = \begin{bmatrix}[B_f][L][C]\\ [B_f][L][S]\end{bmatrix}\{q\}_2\\ &\quad + 2\begin{bmatrix}[B_f][S]\\ [B_f][C]\end{bmatrix}\{q\}_{cc}.\end{aligned} \tag{5.107}$$

We denote:

$$[Q_2] = \begin{bmatrix} [B_f][L][C] \\ [B_f][L][S] \end{bmatrix}; \quad [Q_c] = \begin{bmatrix} [B_f][S] \\ [B_f][C] \end{bmatrix}; \tag{5.108}$$

$$\{\ddot{q}\} = [\ddot{q}_1 \ \ddot{q}_2 \ \ldots \ \ddot{q}_{2x(s-r)} \ \ddot{q}_{2x(s-r)+1} \ \ldots \ \ddot{q}_1 \ ]^T = \begin{Bmatrix} \{\ddot{q}\}_n \\ \cdots \\ \{\ddot{q}\}_c \end{Bmatrix}. \tag{5.109}$$

One obtains:

$$[P]\{\ddot{q}\} = [Q_2]\{q\}_2 + 2[Q_c]\{q\}_c; \tag{5.110}$$

or:

$$[P]_n\{\ddot{q}\}_n + [P]_c\{\ddot{q}\}_c = [Q_2]\{q\}_2 + 2[Q_c]\{q\}_c; \tag{5.111}$$

and the unknown acceleration vector results:

$$\{\ddot{q}\}_n = -[P]_n^{-1}[P]_c\{\ddot{q}\}_c + [P]_n^{-1}\left([Q_2]\{q\}_2 + 2[Q_c]\{q\}_c\right). \tag{5.112}$$

The representation of the solutions of the equations to determine the velocity and accelerations in the form (5.102) and (5.112) allows a numerical approach to the problem of kinematic determination of planar mechanisms.

References

1. Morin D (2008) Introduction to classical mechanics: with problems and solutions, 1st edn. Cambridge University Press
2. Taylor JR, Adam J (2020) Classical mechanics student solutions manual. University Science Books
3. Helliwell TM, Sahakian VV (2020) Modern classical mechanics. Cambridge University Press
4. Rădoi M, Deciu E, Voiculescu, D (1974) Mechanics. Statics and kinematics. Polytechnic Institute Bucharest
5. Voinea R, Atanasiu M (1964) New analytical methods in the theory of mechanisms. Technical Ed., Bucharest.
6. Old Tool Heaven (2026) Millers Falls history. Available at: https://oldtoolheaven.com/millers-falls/history/history4.htm. Accessed 5 Feb 2026
7. Wikipedia (2026) Tîrbuşon. Available at: https://ro.wikipedia.org/wiki/Tirbu%C8%99on. Accessed 5 Feb 2026
8. ISO Republic (2026) Books on table image. Available at: https://isorepublic.com/photo/books-on-table. Accessed 5 Feb 2026
9. Vlase S, Negrean I, Marin M, Năstac S (2020) Kane's method-based simulation and modeling robots with elastic elements, using finite element method. Mathematics 8(5):805. https://doi.org/10.3390/math8050805
10. Stanciu MD, Vlase S, Marin M (2019) Vibration analysis of a guitar considered as symmetrical mechanical system. Symmetry 11(6):727. https://doi.org/10.3390/sym11060727

11. Scutaru ML, Vlase S, Marin M (2023) Symmetrical mechanical system properties-based forced vibration analysis. J Comput Appl Mech 54(4):501–514. https://doi.org/10.22059/JCAMECH.2023.366509.890
12. Scutaru ML, Vlase S (2012) Some properties of motion equations describing the nonlinear dynamical response of a multibody system with flexible elements. J Appl Math 1:628503. https://doi.org/10.1155/2012/628503
13. Vlase S, Teodorescu PP (2013) Elasto-dynamics of a solid with a general 'rigid' motion using FEM model. Part I. Theoretical approach. Roman J Phys 58(7–8):872–881
14. Vlase S, Marin M, Öchsner A (2019) Considerations of the transverse vibration of a mechanical system with two identical bars. Proc Inst Mech Eng Part L: J Mater: Des Appl 233(7):1318–1323. https://doi.org/10.1177/1464420717745109
15. Marin M (1996) Generalized solutions in elasticity of micropolar bodies with voids. Revista de la Academia Canaria de Ciencias 8(1):101–106. https://dialnet.unirioja.es/servlet/articulo?codigo=2899045
16. Marin M (1998) Contributions on uniqueness in thermoelastodynamics on bodies with voids. Cienc Mat (Havana) 16(2):101–109
17. Stanciu E, Bencze A, Cazan A, Vlase S, Marin M, Öchsner A (2025) Effects of infill parameter variations on mechanical properties in 3D printed structures: comparative study with FEM analysis. J Comput Appl Mech 56(2):318–330. https://doi.org/10.22059/jcamech.2025.390563.1371

Chapter 6
Technical Applications of Kinematics

6.1 Motion Transmission

6.1.1 Transmissions with Flexible Elements

(a) Transmissions Between Wheels with Parallel Axles.

Belts or other flexible elements (cables, wire) are used to transmit movement, generally rotational, between two bodies located at a distance. As a general rule, they are used to transmit a rotational movement from one wheel to another, with axes far apart and oriented differently. Belts had a spectacular development in the first period of industrial development, when working speeds and transmitted forces were relatively low. They had the advantage of being reliable, of being able to transmit movement between axes located at a significant distance and with different orientation, and of being sufficiently flexible. In addition, they also played the role of a limiting coupling, when a certain force was exceeded, they began to slide, protecting the system from overloads. In Fig. 6.1, three examples of motion transmission through belts are shown: in Fig. 6.1a with keeping the direction of rotation; in Fig. 6.1b with changing the direction of rotation and in Fig. 6.1c is presented a transmision where the direction of motion is changed.

There exists a rich literature studying different types of technical applications [1–10]. To illustrate the domain industrial application are presented in Figs. 6.2, 6.3 and 6.4.

If there is no slip between the belt and the wheels, then the speed of all points of the belt will be equal to the peripheral speed of the wheels, so:

$$v = \omega_1 R_1 = \omega_2 R_2, \tag{6.1}$$

in both cases. The transmission ratio is defined as the ratio between the angular velocity of the driving wheel and the angular velocity of the driven wheel:

S. Vlase et al., *Models in Kinematics for Engineers*,
https://doi.org/10.1007/978-3-032-28113-5_6

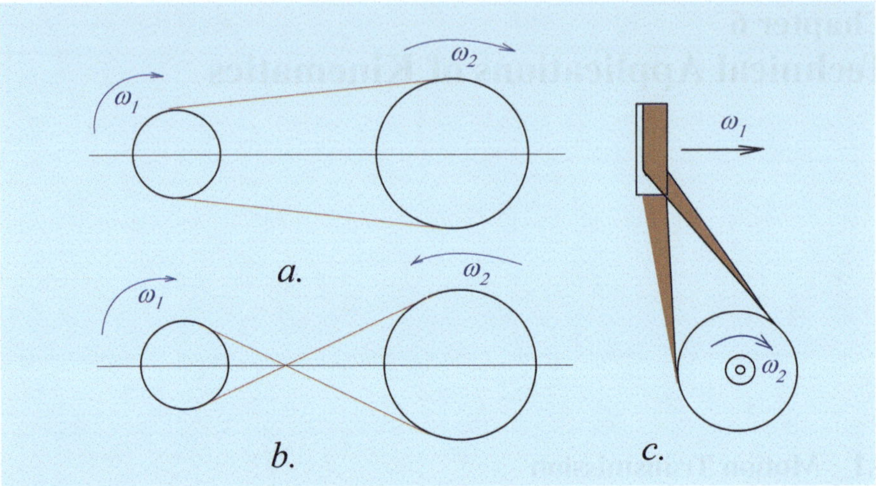

Fig. 6.1 Transmission through belts: **a** The direction of rotation is keeped; **b** The direction is reversed; **c** The direction is changed

Fig. 6.2 Transmission through belts. Industrial applications (photo authors)

Fig. 6.3 Belt-driven lathes (photo authors)

Fig. 6.4 Chain transmissions (Friedmann-Knoller steam car from 1904, displayed at the Vienna Technical Museum, photo: authors)

$$i_{12} = \frac{\omega_1}{\omega_2} = \frac{n_1}{n_2}, \tag{6.2}$$

where from:

$$\omega_2 = \omega_1 / i_{12}. \tag{6.3}$$

Belt transmissions have the disadvantage of slippage, so the torque that can be transmitted is limited to a value determined by slippage.

(b) Transmissions Between Wheels with Perpendicular Axes
Figure 6.1c shows such a transmission. The calculation relationships remain those from Eq. (6.1), considering that all points of the belt have the same speed.
(c) Chain Transmissions
It is treated in the same way as belt drives (Fig. 6.4). Chain transmissions avoid slippage between the transmission element and the wheel. They were widely used to transmit the moment from one axis to another and have the advantage that they do not allow sliding.

At the moment, they have been largely replaced by toothed belts, which take the advantages of chain transmission, are more reliable, quieter and have a reasonable cost price.

(d) Variators with Belts
It ensures a variable transmission ratio of the torque continuously, or in steps, depending on the chosen construction option.

Figure 6.5a shows a variator with continuously changing speed and in Fig. 6.5b shows a speed variator with belts with a step change in speed.
Figure 6.6 presents the applications in the industry (lathes).

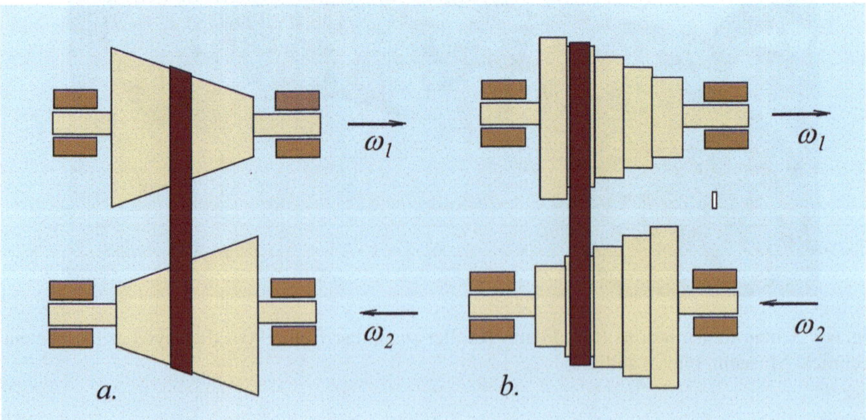

Fig. 6.5 *a* Variator with continuously changing speed; *b* speed variator with a step change in speed

Fig. 6.6 Variator with belts on a lathe (photo authors)

6.1.2 Transmissions with Friction Wheels

6.1.2.1 Friction Cylinders

We consider two cylinders (Δ_1) with (Δ_2) parallel axes. The surfaces in contact are the cylinders of revolution, the contact being made after the generators (Figs. 6.7 and 6.8).

Fig. 6.7 Friction wheel with parallel axes [11]

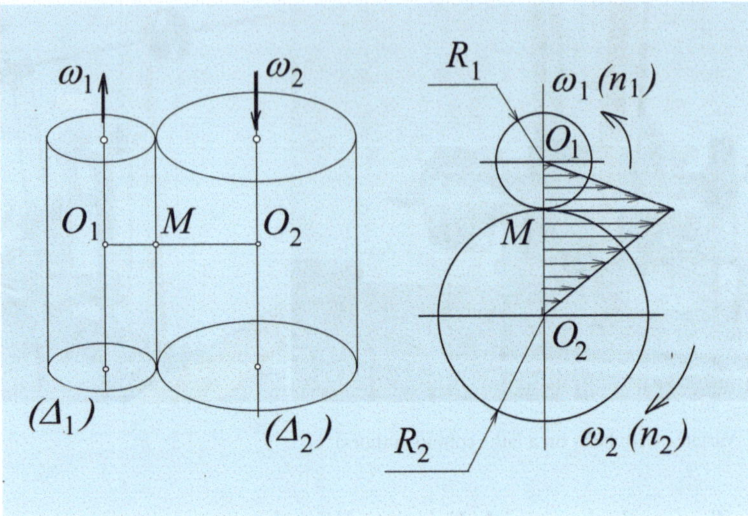

Fig. 6.8 Calculus for friction wheel with parallel axes

The condition that the two surfaces do not slide against each other is:

$$\vec{v}_{1M} = \vec{v}_{2M}, \tag{6.4}$$

or:

$$\vec{\omega}_1 \times \overrightarrow{O_1M} = \omega_2 \times \overrightarrow{O_2M}, \tag{6.5}$$

from where, taking into account that: $\vec{\omega}_1 \perp \overrightarrow{O_1M}$ and $\vec{\omega}_2 \perp \overrightarrow{O_2M}$, we obtain:

$$\frac{O_1M}{O_2M} = \frac{\omega_2}{\omega_1} = ct = \frac{R_1}{R_2}. \tag{6.6}$$

The relationship therefore results:

$$\omega_1 R_1 = \omega_2 R_2, \tag{6.7}$$

used in solving problems of this kind. If the angular velocity ω is measured in rad/s and speed n in RPM, then we have:

$$\omega = \frac{\pi n}{30}. \tag{6.8}$$

So we can also write:

$$n_1 R_1 = n_2 R_2. \tag{6.9}$$

6.1.2.2 Friction Cones

By hypothesis, the axes (Δ_1) and (Δ_2) are concurrent in S and make the angle β between them. The contact surfaces are cones of revolution (Figs. 6.9 and 6.10). The no-slip rolling condition is:

$$\vec{v}_{1M} = \vec{v}_{2M}, \tag{6.10}$$

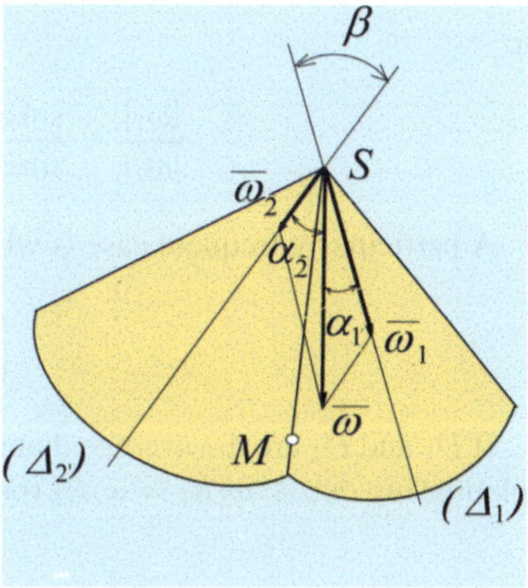

Fig. 6.9 Concurent axes

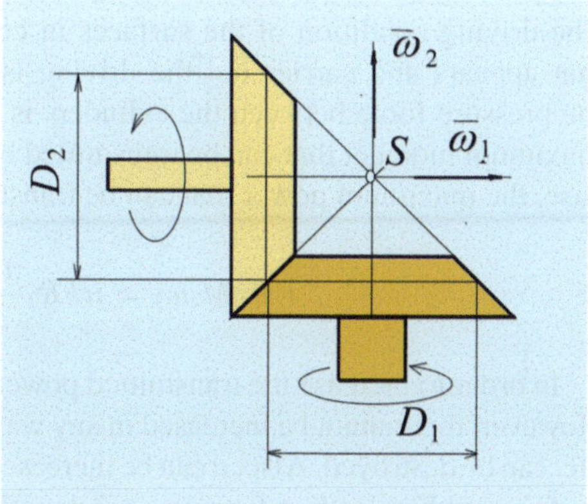

Fig. 6.10 Perpendicular axes

M being a point of contact. Or:

$$\vec{\omega}_1 \times \overrightarrow{SM} = \omega_2 \times \overrightarrow{SM}, \tag{6.11}$$

or:

$$(\vec{\omega}_2 - \vec{\omega}_1) \times \overrightarrow{SM} = 0. \tag{6.12}$$

If it is written: $\vec{\omega} = \vec{\omega}_2 - \vec{\omega}_1$ it results: $\vec{\omega} \times \overrightarrow{SM} = 0$, from which it follows that the point M is located along the relative angular velocity $\vec{\omega}$. The contact surfaces are two cones of revolution with axes (Δ_1) and (Δ_2) having semi-angles α_1 and α_2. From the triangle *SAB* it follows:

$$\frac{|\omega_1|}{\sin \alpha_2} = \frac{|\omega_2|}{\sin \alpha_1}, \tag{6.13}$$

or:

$$\frac{|\omega_1|}{|\omega_2|} = \frac{\sin \alpha_2}{\sin \alpha_1} = \frac{n_1}{n_2}, \alpha_1 + \alpha_2 = \beta. \tag{6.14}$$

A particularly frequent case is when $\alpha_1 + \alpha_2 = \pi/2$ (Fig. 6.10). Then:

$$\frac{\omega_2}{\omega_1} = tg\alpha_1. \tag{6.15}$$

If D_1 and D_2 are the average diameters of the truncated cones in contact, then the relationship exists: $\omega_1 R_1 = \omega_2 R_2$ (or $n_1 R_1 = n_2 R_2$), where $D_1 = 2R_1, D_2 = 2R_2$.

6.1.2.3 Operating Conditions

The driving condition of the surfaces in contact requires that the tangential force that appears and carries out the driving is lower than the limit friction force. If the pressure force between the cylinders is F, the coefficient of friction μ, then the maximum moment that can be transmitted by wheel 2 is $M_2 = TR_2 = \mu FR_2$. In this case, the maximum power that can be transmitted to wheel 2 is (Fig. 6.11).

$$P = M_2\omega_2 = \mu FR_2 \frac{\pi n_2}{30} = \mu F \frac{\pi n_1 R_1}{30}. \tag{6.16}$$

In order to increase the transmitted power, we must increase the pressing force F. However, this cannot be increased in any way because the friction surfaces, bearings, etc. can be destroyed. Also, μ can be increased by covering the cylinder 1 with a rough surface (rubber, leather, ferrodou,...) the other cylinder being cast iron.

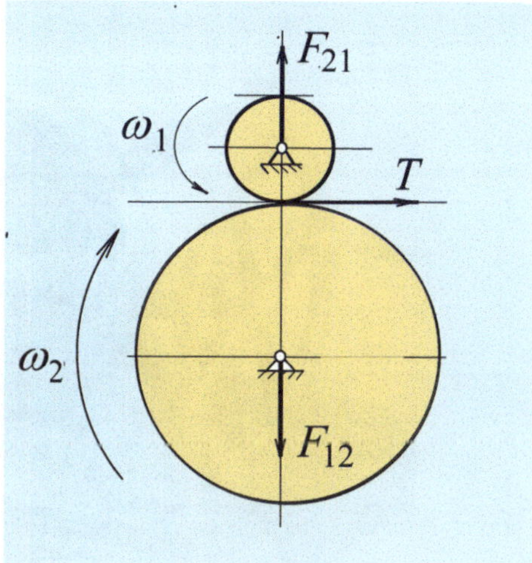

Fig. 6.11 Calculation of the capable force

Values of μ:

- Cast iron on skin: 0.0.2–0.3
- Cast iron on ferrodou: 0.25
- Cast iron on wood: 0.0.2–0.4

A friction wheel drive is used in cases where the power to be transmitted is small. For example speed reducers of low power electric motors, friction presses, speed variators, clothes wringers, motorized bicycles (Fig. 6.12).

The advantages of friction wheels would be the quick realization with low costs, the silent operation, they can be used as torque limiters.

Disadvantages are: Slippage operation (< 10%) especially at start-up, rapid wear of friction surfaces, bending of shafts, power is limited by grip.

Example: Speed variator (friction plate and roller) (Fig. 6.13). The axes (Δ_1) and (Δ_2) are perpendicular. The variator allows obtaining a progressive change in speed

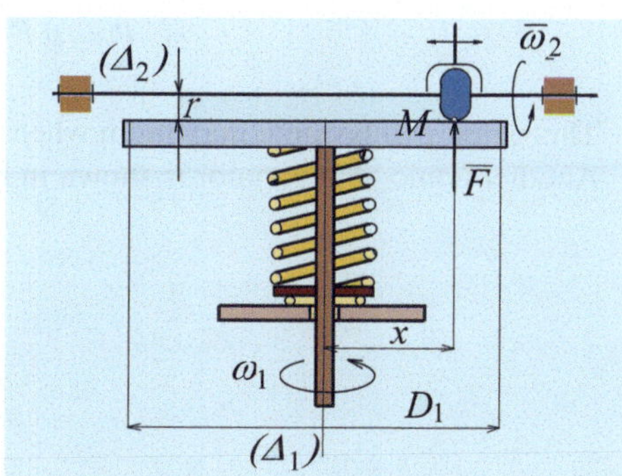

Fig. 6.12 Variator with friction wheels (perpendicular axes)

Fig. 6.13 Friction wheels with perpendicular axes [12]

by moving the bucket along the plate. The plate rotates with the angular velocity ω_1 (n_1). The spring exerts a force F and the coefficient of friction between the roller and the plate is μ. If x is the distance at which the bucket is located, the condition for transmission of motion by friction gives: $n_1 x = n_2 r$ it results. If $x > r$, it results $n_2 > n_1$ (speed multiplier) and if $x < r$, it results $n_2 < n_1$ (speed reducer). If $x = 0$, then $n_2 = 0$. The maximum moment that can be transmitted is:

$$M = Tr = \mu F r, \tag{6.17}$$

and the maximum power that can be transmitted when the bucket is at distance x:

$$P = \mu F \frac{\pi n x}{30}. \tag{6.18}$$

This value will become maximum when $x = R_1$.

Another conic gear variator is shown in Fig. 6.14.

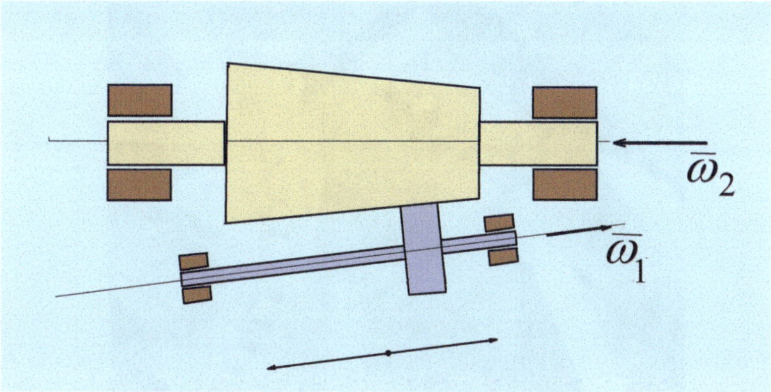

Fig. 6.14 Speed variator

6.1.3 *Transmission of Motion Through Gears*

Generalities on Gears

Friction wheels have the advantage of being easy and inexpensive to make, with quiet operation that can also be used as torque limiters. At the same time, it also presents disadvantages, which are: Sliding, especially at start-up, rapid wear of the surfaces, bending of the shafts due to the pressing force that must be achieved, the transmitted power limited by sliding. To remove some of the inconveniences, teeth are used in the wheels in contact, making the drive through an obstacle. From a historical point of view, gearing is a very old process, found, for example, in village mills (Fig. 6.15).

The assembly made up of two toothed wheels represents a gear. The gears allow the transmission of circular motion between two neighboring shafts, the gearing being done by pressing a tooth of the driving wheel against a tooth of the driven wheel. For certain tooth profiles, the transmission ratio can be rigorously constant and the transmitted torque is limited only by the resistance of the tooth, which is generally very high.

Figures 6.16 and 6.17 show examples of cylindrical gears with straight teeth and Fig. 6.18 a historical representation of a gear.

Teeth with Involute Profile

In the following we will show the advantages of the most widespread profile in technique, which is the involute one. These are given by the fact that the transmission ratio is rigorously constant and that there is no slippage in the motion of one tooth relative to another. We assume that the circles (C_1) and (C_2) are tangent in I_{21}. Let us consider two circles (C_1') and (C_2') of radii R_1' and R_2'. The line T_1T_2 tangent to the circles (C_1) and (C_2) can be considered as rolling without slipping on the two moving circles. The point K belonging to the line T_1T_2 describes an involute if we refer to the first circle and $arcT_1B_1' = T_1K$. The two obtained involutes represent a

Fig. 6.15 Cylindrical gears in a windmill [13]

Fig. 6.16 Cylidrical gears with straight teeth (photo authors)

Fig. 6.17 An industrial application of cylindrical gears (photo authors)

conjugate profile, they are tangent in K and their common normal passes through I_{21}. The contact point K is on the line T_1T_2. During the rotation of the two circles, the line is motionless, point K is fixed and only the two involutes move, remaining permanently in contact and tangents. The angle α in the figure is called the pressure angle (Fig. 6.19).

Transmission Ratio. During the motion the point B_1 becomes B_1' and B_2 becomes B_2'. Taking into account the involute study in Chap. 1, it follows that:

$$arcB_1B_1' = I_{21}K = arcB_2B_2'. \tag{6.19}$$

If ω_1 and ω_2 are respectively the angular velocities of the two circles, it can be written:

$$\omega_1 = \frac{arcB_1B_1'}{R_1'\Delta t}; \omega_2 = \frac{arcB_2B_2'}{R_2'\Delta t}, \tag{6.20}$$

where from:

$$\frac{\omega_2}{\omega_1} = \frac{R_1'}{R_2'}. \tag{6.21}$$

Fig. 6.18 Raising a slab of stone using a block-and-tackle mechanism [14]

From the similarity of the triangles $O_1 I_{21} T_1$ and $O_2 I_{21} T_2$, it follows immediately:

$$\frac{\omega_2}{\omega_1} = \frac{R_1'}{R_2'} = \frac{R_1}{R_2}, \tag{6.22}$$

which shows that we have sliding without rolling of circles (C_1) and (C_2) (Fig. 6.20).

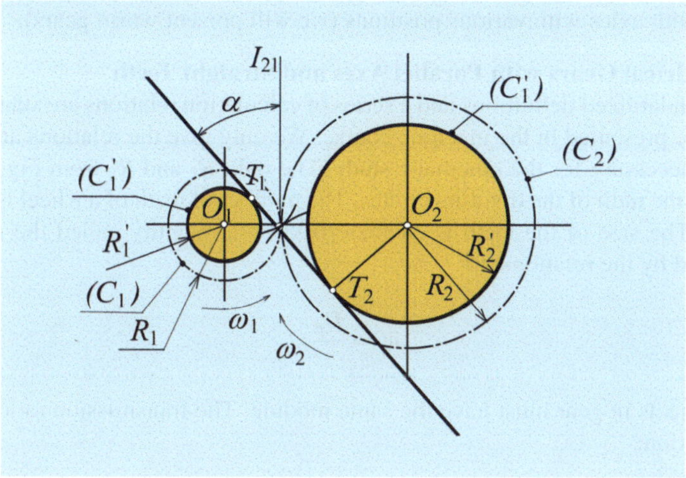

Fig. 6.19 The study of involute profile gearing

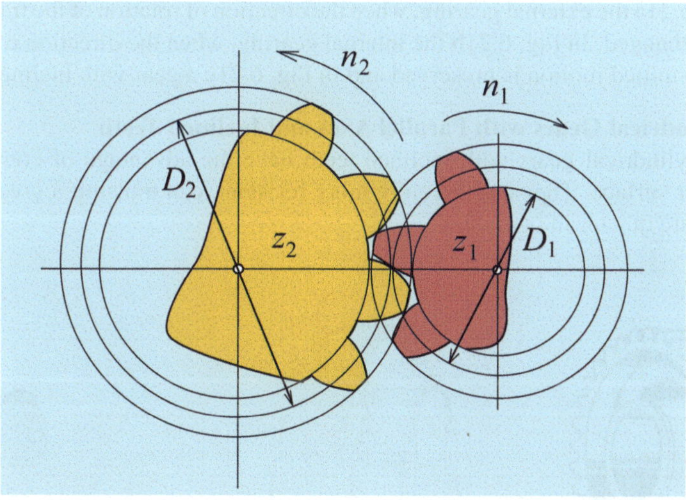

Fig. 6.20 Gear

A Classification of Gears

According to the relative position of the axles of the wheels in gearing, the following classification can be made:

- Gears with parallel axes (we will present gears with straight teeth and inclined teeth);
- Gears with concurrent axes (we will present conic gears with straight teeth);

- Gears with axles with various positions (we will present worm gears).

(i) **Cylindrical Gears with Parallel Axes and Straight Teeth**

Standardized definitions and a series of calculation relations are standardized and are presented in the machine course. We only give the relations and definitions necessary for the kinematic study. The radii R_1 and R_2 from Fig. 6.19 are called the radii of the dividing circles. The number of teeth of a wheel is denoted by z. The size of the teeth is characterized by a quantity called the modulus, defined by the relationship:

$$m = \frac{D_p}{z}. \tag{6.23}$$

Two wheels in gear must have the same module. The transmission ratio is given by the relation:

$$i_{2,1} = \frac{\omega_2}{\omega_1} = \frac{n_2}{n_1} = \frac{R_1}{R_2} = \frac{2mz_1}{2mz_2} = \frac{z_1}{z_2}. \tag{6.24}$$

In Fig. 6.21a the external gearing, when the direction of rotation of the transmitted motion is changed. In Fig. 6.21b the internal gearing, when the direction of rotation of the transmitted motion is preserved and in Fig. 6.21c a gear with inclined teeth.

(ii) **Cylindrical Gears with Parallel Axes and Inclined Teeth**

Cylindrical gears with inclined teeth have the advantage of contact on a larger surface. They are therefore more resistant, can transmit a greater force and are quieter in operation.

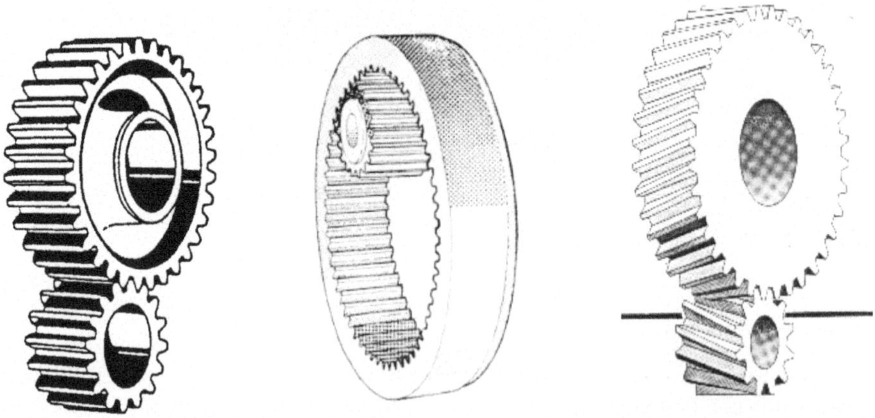

a. b. c.

Fig. 6.21 Gear with parallel axes: **a** external gear; **b** internal gear; **c** parallel axes with inclined teeth

Fig. 6.22 Worm gear [15]

(iii) **Worm Gear**

The rotational motion is transmitted between two perpendicular axes and a significant reduction in the transmitted speed can be obtained. The transmission of the motion is done from the endless screw to the shaft driven to the worm wheels. The endless screw is also called a worm, hence the name worm-worm gear. If it is considered that the screw has only one start and the worm wheel has z teeth, then an empty screw thread is equal to a full worm wheel, so they have the same pitch. During the time when the screw makes a complete rotation, i.e. it rotates by 2π radians, the wheel rotates by one tooth, i.e. by $2\pi/z$ rad. So at a rotation of the screw angle α, the worm wheel will rotate with the angle β given by $\beta = \alpha/z$ Fig. 6.22.

It follows that the ratio between the angular velocities and the transmission ratio is:

$$i_{12} = \frac{\omega_1}{\omega_2} = z. \tag{6.25}$$

If the screw has n_f starts, then at one complete rotation of it (2π radians) the worm wheel will rotate with $n_f = 2\pi/z$ rad, so the wheel will advance with n_f teeth, corresponding to the n_f threads cut in one pitch of the screw with a single lead. In this case the transmission report will be given by the relationship:

$$i_{12} = \frac{\omega_1}{\omega_2} = \frac{z}{n_f}. \tag{6.26}$$

Fig. 6.23 Rack [16]

(iv) **Rack**

The rack (Fig. 6.23) can be assimilated to a wheel with infinite radius. It is in contact with a toothed wheel of the same module and the rotational motion is transformed into translational movement or vice versa. The relationship between the angular speed of the wheel and the translational speed of the rack is: $v = \omega \cdot R$.

(v) **Transmissions with Conical Gears with Straight Teeth**

In this case, the transmission of the motion is done through cones that have applied to them gear (Figs. 6.24 and 6.25). In this case, the considerations made for transmission through conical friction wheels and the relationships obtained there remain valid. In the transmission report, the radii can be replaced by the numbers of teeth.

Gears have many uses in practice, especially in motor vehicles. For example, they are used for gearboxes, reducers, normal and interaxial differentials, for transmitting engine rotation to different components of the vehicle (Figs. 6.26, 6.27, 6.28 and 6.29)

6.1.4 Cam Mechanisms

Cam mechanisms are of great technical importance (Fig. 6.30). They have the role of transforming motion. Figures 6.31, 6.32, 6.33, 6.34 and 6.35 show a series of cam mechanisms in which the driving element is called cam and the element that follows the motion is called tappet or pusher.

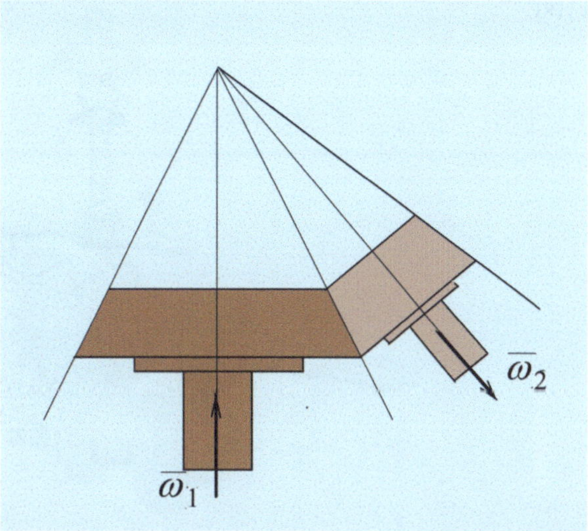

Fig. 6.24 Schematic of a conic gear

Fig. 6.25 Conic gear [17]

The cam has a profiled shape and transmits to the driven element (usually a rod) a certain motion whose law is determined by the profile of the cam.

A desired motion law will determine the cam profile that is obtained through synthesis. The kinematic and operating conditions in the case of cam mechanisms

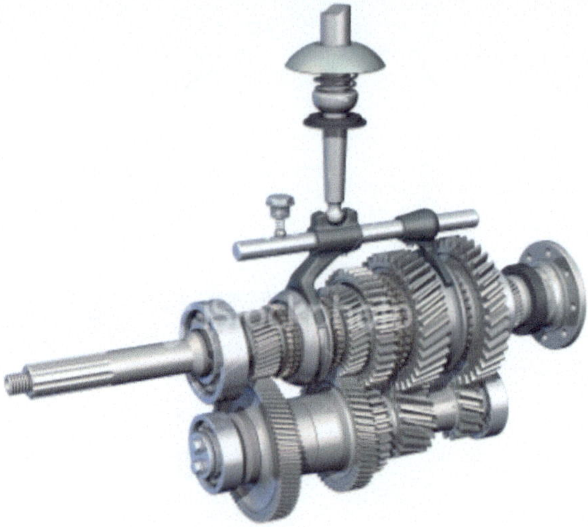

Fig. 6.26 Gearbox [18]

Fig. 6.27 Reducer [19]

are, in general, difficult to describe. The difficulties are due to the geometry of such systems.

Figure 6.31 shows the transformation of translational motion into rotational motion, in Fig. 6.32 the transformation of rotational motion into rotational motion and in Figs. 6.33 and 6.34 the transformation of rotational motion into motion of translation.

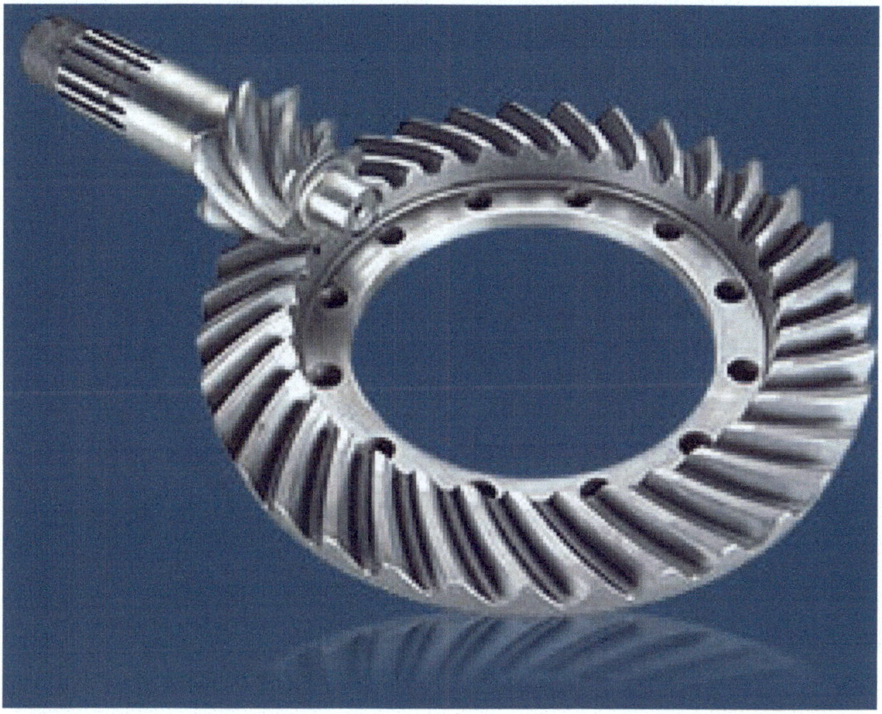

Fig. 6.28 The toothed crown of a differential [20]

Figure 6.35a and b show the transformation of rotational motion into translation and Fig. 6.35c the transformation of rotational motion into alternating rotational motion.

6.1.5 Cardan Transmissions

Cardan transmissions are used for transmission of rotational motion, without amplifying the transmitted torque (on average). At the moment, the cardan transmission means an assembly of machine parts (joints, shafts, safety couplings, shock absorbers, intermediate bearings, etc.) that constitute an independent functional unit that serves to transmit mechanical energy at a distance, through rotational movement, without amplifying the average torque (Fig. 6.36).

The transmission law for a cardan mechanism as in Fig. 6.36 is given by:

$$tg\phi_2 = \frac{tg\phi_1}{\cos\alpha}. \tag{6.27}$$

Fig. 6.29 Planetary with reducer role in the wheel [20]

Fig. 6.30 A cam and follower mechanism [21]

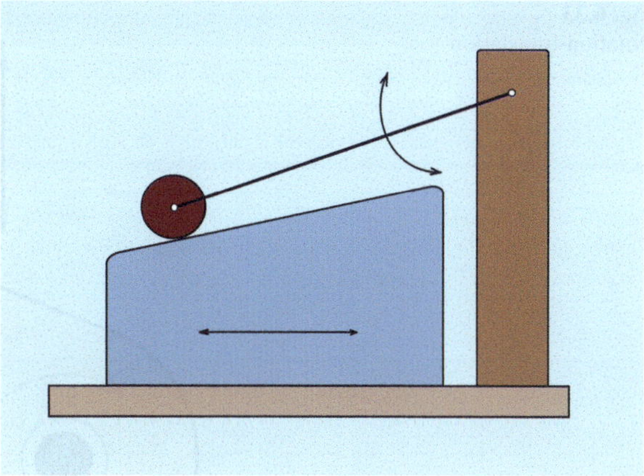

Fig. 6.31 Translation-rotation

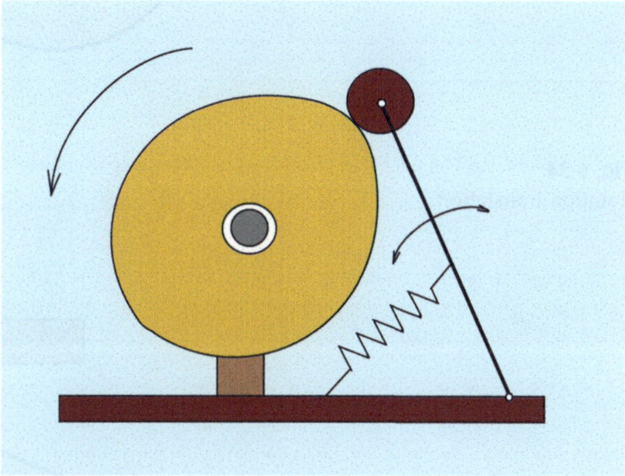

Fig. 6.32 Rotation-rotation

deduced using spherical trigonometry [22].

Some version of a Cardan transmission is presented in Fig. 6.37.

A documented presentation on the use of cardan can also be found in [22] and an application in Fig. 6.38 [23].

6.2 The Study of Simple Mechanisms

The present paragraph aims to study some simple mechanisms, often used in the construction of machines. The quadrilateral mechanism and the connecting rod-crank (crank-slider) mechanism will be studied (Fig. 6.39).

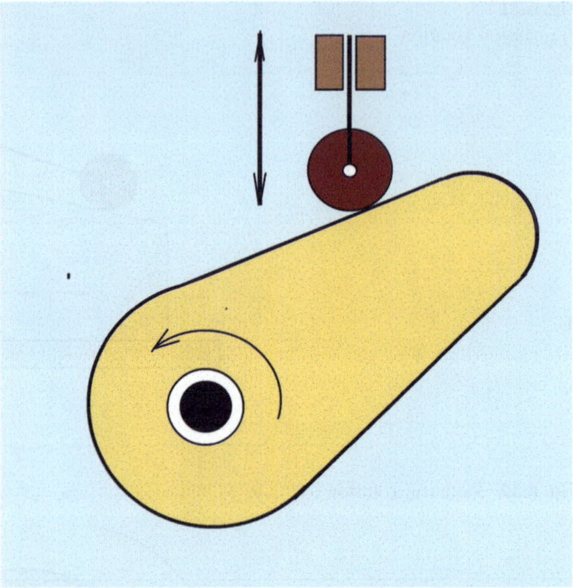

Fig. 6.33
Rotation-translation

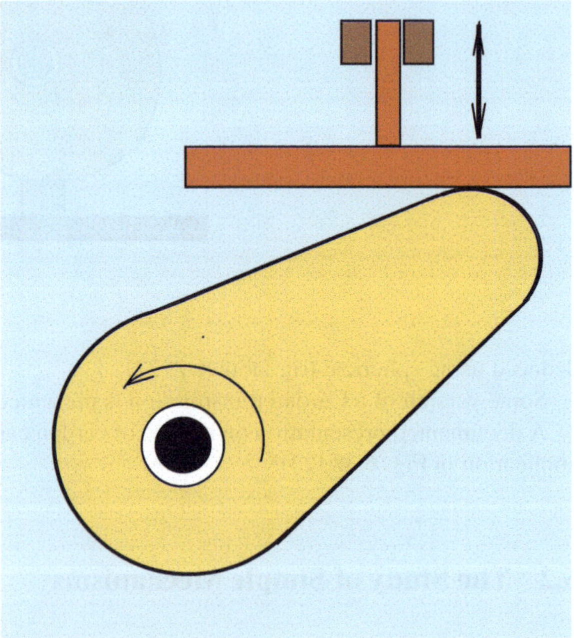

Fig. 6.34
Rotation-translation

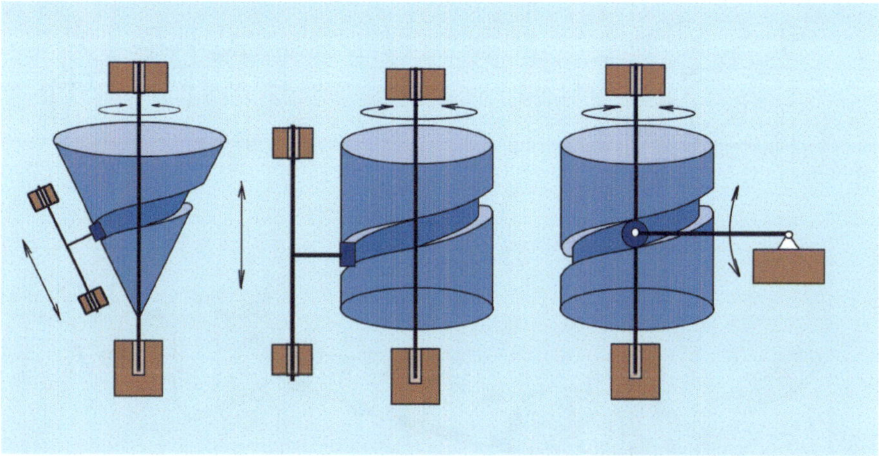

Fig. 6.35 Types of mecanism cam and follower; **a** Rotation-translation; **b** Rotation-translation; **c** Translation-translation

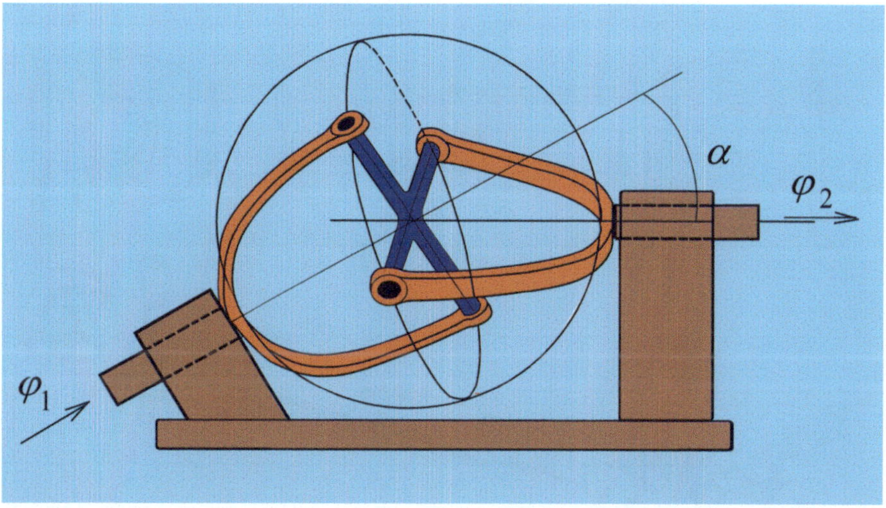

Fig. 6.36 Cardan transmission

6.2.1 Quadrilateral Mechanism

To solve the mechanism (Fig. 6.40) the method of loop-cloosing equations will be applied. The mechanism is defined by a single contour and the closure equation for the mechanism will be:

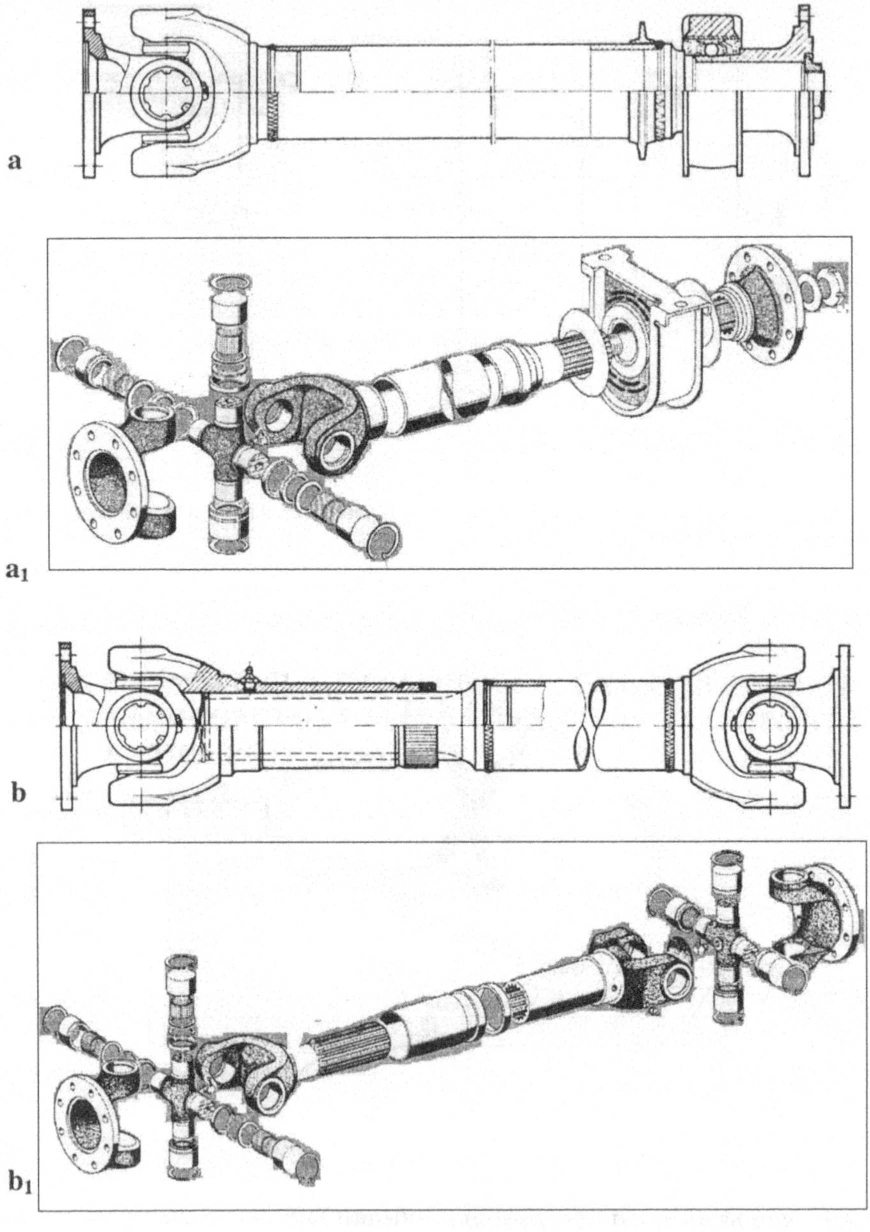

Fig. 6.37 Constructive variants of cardan transmissions. **a** Monocardan shaft; **b** bicardanic shaft [22]

Fig. 6.38 Cardan mechanism for transporting the wounded person [23]

$$\overrightarrow{AB} + \overrightarrow{BC} + \overrightarrow{CD} + \overrightarrow{DA} = 0. \tag{6.28}$$

If the equation is projected onto the two axes of the reference system, with the notations in the figure, we obtain:

$$l_1 \cos\phi_1 + l_2 \cos\phi_2 + l_3 \cos\phi_3 = l_4$$

$$l_1 \sin\phi_1 + l_2 \sin\phi_2 + l_3 \sin\phi_3 = 0. \tag{6.29}$$

The two relations should provide the angles ϕ_2 and ϕ_3 as a function of the lengths of the bars and the angle of the leading element ϕ_1. The problem is quite difficult to solve analytically, but at the moment we have sufficiently powerful mathematical software packages to ensure our solution. In the case of an analytical approach, the triangles *ABD* and *BCD* are solved successively.

If it is written with $\omega_1 = \dot{\phi}_1; \omega_2 = \dot{\phi}_2; \omega_3 = \dot{\phi}_3$ the angular velocities of the respective bars, the relations are obtained by derivation:

$$\begin{aligned} \omega_1 l_1 \sin\phi_1 + \omega_2 l_2 \sin\phi_2 + \omega_3 l_3 \sin\phi_3 = 0; \\ \omega_1 l_1 \cos\phi_1 + \omega_2 l_2 \cos\phi_2 + \omega_3 l_3 \cos\phi_3 = 0, \end{aligned} \tag{6.30}$$

Fig. 6.39 Four bars mechanism in industry (photo authors)

from which the angular velocities of bars *BC* and *CD* result:

$$\omega_2 = \omega_1 \frac{l_1 \sin(\phi_1 - \phi_3)}{l_2 \sin(\phi_3 - \phi_2)}; \ \omega_3 = \omega_1 \frac{l_1 \sin(\phi_1 - \phi_2)}{l_3 \sin(\phi_3 - \phi_2)}. \tag{6.32}$$

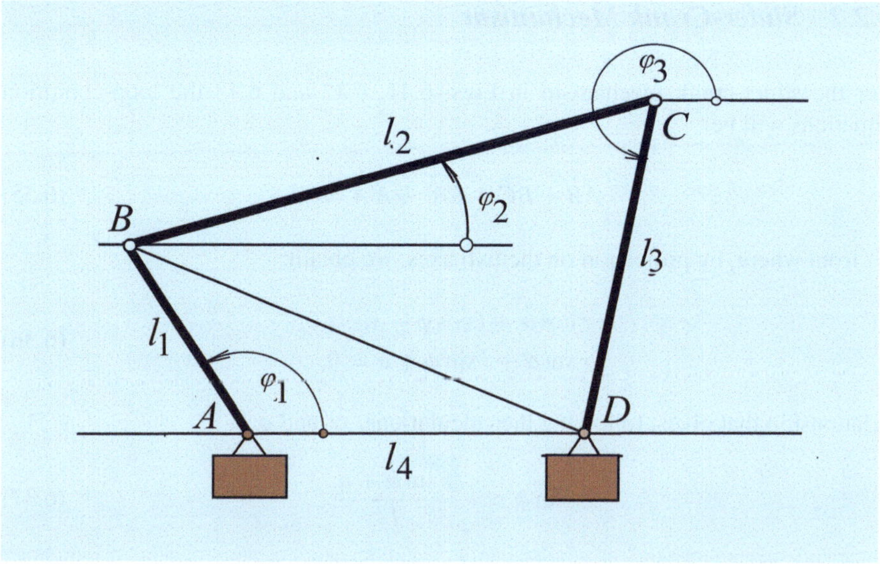

Fig. 6.40 Schetch of a four bars mechanism

By derivation we obtain the relations that give us the angular accelerations of the bars:

$$\varepsilon_1 l_1 \sin\phi_1 + \omega_1^2 l_1 \cos\phi_1 + \varepsilon_2 l_2 \sin\phi_2 + \omega_2^2 l_2 \cos\phi_2 + \varepsilon_3 l_3 \sin\phi_3 + \omega_3^2 l_3 \cos\phi_3 = 0$$
$$\varepsilon_1 l_1 \cos\phi_1 - \omega_1^2 l_1 \cos\phi_1 + \varepsilon_2 l_2 \cos\phi_2 - \omega_2^2 l_2 \sin\phi_2 + \varepsilon_3 l_3 \cos\phi_3 - \omega_3^2 l_3 \sin\phi_3 = 0, \tag{6.33}$$

from where, by calculation, we obtain ε_1 and ε_2. We have:

$$\begin{bmatrix} l_2 \sin\phi_2 & l_3 \sin\phi_3 \\ l_2 \cos\phi_2 & l_3 \cos\phi_3 \end{bmatrix} \begin{Bmatrix} \varepsilon_2 \\ \varepsilon_3 \end{Bmatrix} = -l_1 \varepsilon_1 \begin{Bmatrix} \sin\phi_1 \\ \cos\phi_1 \end{Bmatrix} - l_1 \omega_1^2 \begin{Bmatrix} \cos\phi_1 \\ \sin\phi_1 \end{Bmatrix} - l_2 \omega_2^2 \begin{Bmatrix} \cos\phi_2 \\ \sin\phi_2 \end{Bmatrix} - l_3 \omega_3^2 \begin{Bmatrix} \cos\phi_3 \\ \sin\phi_3 \end{Bmatrix}$$
$$= -l_1 \varepsilon_1 \begin{Bmatrix} \sin\phi_1 \\ \cos\phi_1 \end{Bmatrix} - l_1 \omega_1^2 \begin{Bmatrix} \cos\phi_1 \\ \sin\phi_1 \end{Bmatrix} - \omega_1^2 \frac{l_1^2 \sin^2(\phi_1 - \phi_3)}{l_2 \sin^2(\phi_3 - \phi_2)} \begin{Bmatrix} \cos\phi_2 \\ \sin\phi_2 \end{Bmatrix} - \omega_1^2 \frac{l_1^2 \sin^2(\phi_1 - \phi_2)}{l_3 \sin^2(\phi_3 - \phi_2)} \begin{Bmatrix} \cos\phi_3 \\ \sin\phi_3 \end{Bmatrix}. \tag{6.34}$$

6.2.2 Slider-Crank Mechanism

For the slider-crank mechanism in Figs. 6.41, 6.42 and 6.43 the loop-condition equations will be:

$$\overrightarrow{AB} + \overrightarrow{BC} + \overrightarrow{CA'} + \overrightarrow{A'A} = 0, \tag{6.35}$$

from where, by projection on the two axes, we obtain:

$$\begin{aligned} r\cos\alpha + l\cos\phi - x_C &= 0, \\ r\sin\alpha + l\sin\phi + a &= 0, \end{aligned} \tag{6.36}$$

relationship that gives, following the calculations, x_C and φ:

$$\sin\phi = -\frac{r\sin\alpha + a}{l}. \tag{6.37}$$

To determine the angular speed ω_2 and the speed of the slide v_C, the two projections of the loop closing contour equations are derived:

$$\begin{aligned} r\omega_1\sin\alpha + l\omega_2\sin\phi + v_C &= 0, \\ r\omega_1\cos\phi + l\omega_2\cos\phi &= 0. \end{aligned} \tag{6.38}$$

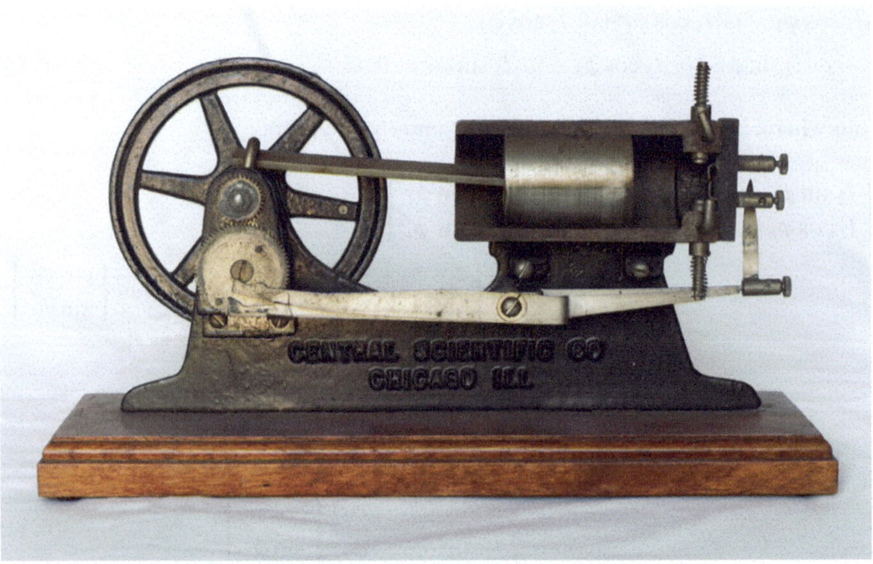

Fig. 6.41 Slider-crank mechanism of an engine [24]

Fig. 6.42 Slider-crank mechanisms in industry (photo authors)

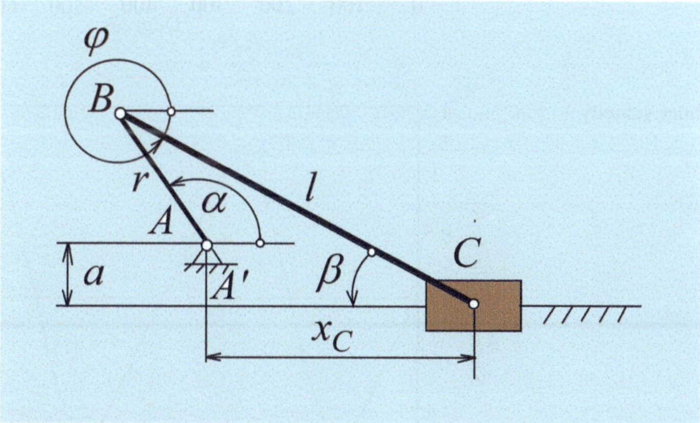

Fig. 6.43 Sketch of the slider-crank mechanism

It results:

$$\omega_2 = -\omega_1 \frac{r\cos\alpha}{l\cos\phi}; \; v_C = r\omega_1\left(-\sin\alpha + \frac{\cos\alpha}{tg\phi}\right). \tag{6.39}$$

The graphs of the angular speed of the connecting rod and the linear speed of the slide are presented in Figs. 6.44 and 6.45, where the angular speed of the crank was considered constant.

Through a new derivation we obtain the relations that give us the accelerations:

$$\begin{aligned} &\omega_1^2 r\cos\alpha + \varepsilon_1 r\sin\alpha + \omega_2^2 l\cos\phi + \varepsilon_2 l\sin\phi + a_C = 0; \\ &-\omega_1^2 r\sin\alpha + \varepsilon_1 r\cos\alpha - \omega_2^2 l\sin\phi + \varepsilon_2 l\cos\phi = 0, \end{aligned} \tag{6.40}$$

Fig. 6.44 Angular speed of the connecting rod

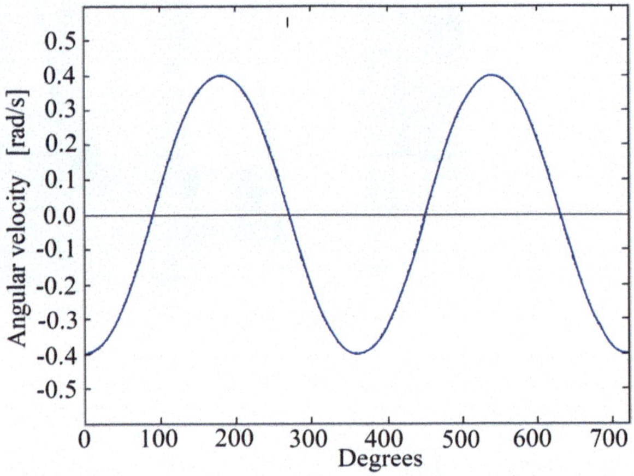

Fig. 6.45 Slider velocity

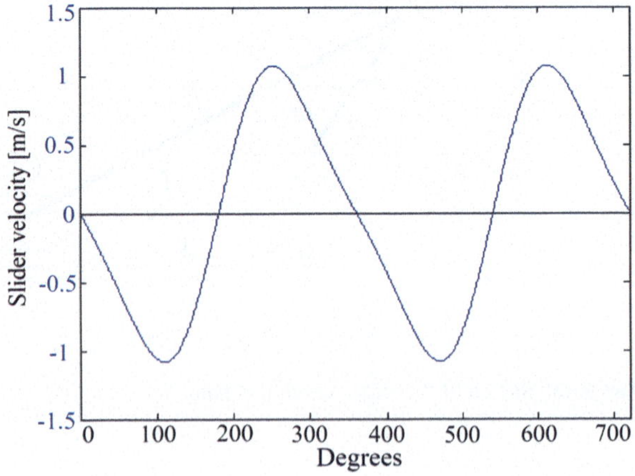

which offers us easily ε_2 and a_C according to known elements:

$$\varepsilon_2 = -\frac{r\cos\alpha}{l\cos\beta}\varepsilon_1 + \left(\frac{r\sin\alpha}{l\cos\beta} - t^2\frac{\sin\beta}{\cos\beta}\right)\omega_1^2 = t\varepsilon_1 + u\omega_1^2, \tag{6.41}$$

$$a_C = (-r\sin\alpha + lt\sin\beta)\varepsilon_1 + (-r\sin\alpha + lt^2\sin\beta + lu\cos\beta)\omega_1^2. \tag{6.42}$$

Careful! If we use the angle β instead of the angle φ in the loop closing equations, then we get $\omega_2 = -\dot{\beta}$ the relation which results easily if we derive $\beta + \phi = 2\pi$ and take into account that $\omega_2 = \dot{\phi}$. Some connecting rod curves for the crank slidder crank mechanism are shown Figs. 6.46, 6.47, 6.48, 6.49.

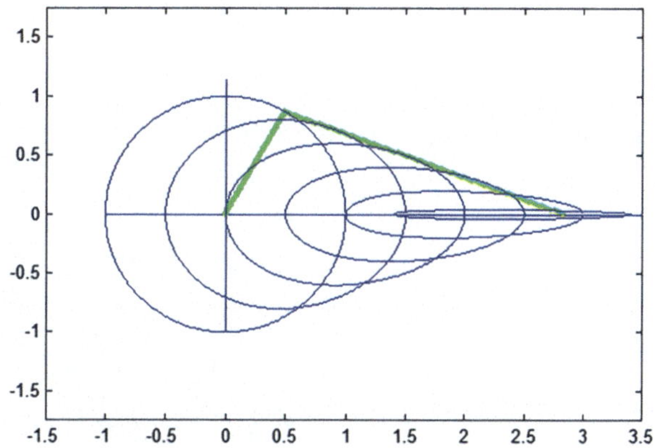

Fig. 6.46 The ratio r/l = 0.4

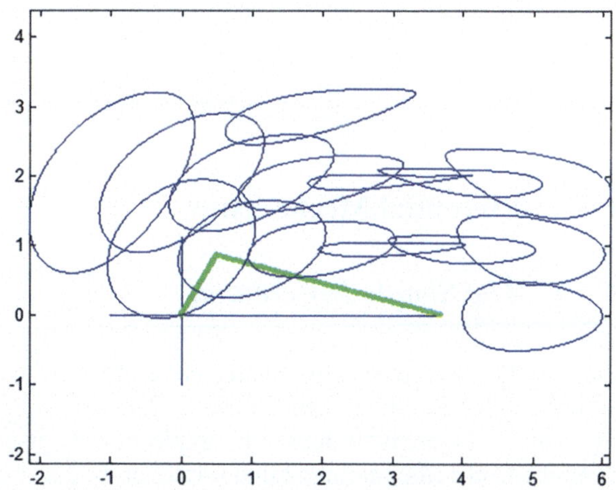

Fig. 6.47 The ratio r/l = 0.2

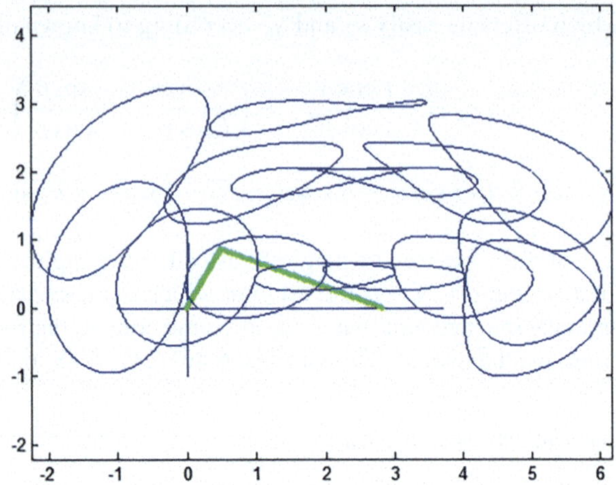

Fig. 6.48 The ratio r/l = 0.4

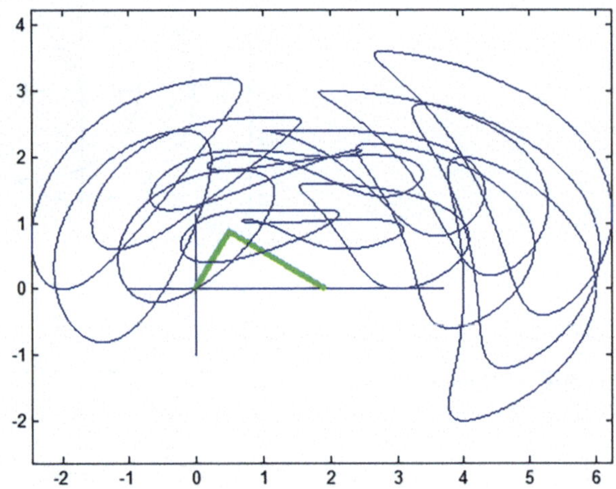

Fig. 6.49 The ratio r/l = 0.6

6.3 Differential Mechanisms

6.3.1 The Simple Differential

Differential mechanisms are characterized by the fact that they have one or more wheels, called satellites, which have mobile axes and rotate around the axis of the mechanism. Figure 6.50 shows the simplest mechanism of this kind consisting of the central wheel 1, the satellite 2 that rotates around an axis fixed in the satellite arm *H* (also called the crank). The satellite arm rotates around the axis of wheel 1.

The described mechanism has two degrees of freedom (two kinematic parameters are needed to describe the motion of the mechanism). In practice these mechanisms

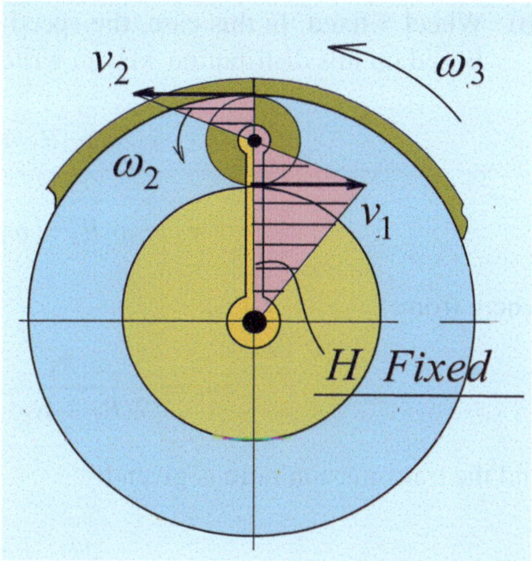

Fig. 6.50 The simple differential

are used as mechanisms with one degree of freedom, obtained from the differential mechanism by blocking an element. We present the possible cases of operation as a mechanism with one degree of freedom (Fig. 6.52).

(a) If we fix the crank, a gear mechanism with fixed axes is obtained (Fig. 6.50). The speed distribution in this case is the one in the figure. Based on this distribution, the relations can be written:

$$\begin{aligned} v_1 &= \omega_1 R_1 = \omega_2 R_2; \\ v_2 &= \omega_2 R_2 = \omega_3 R_3, \end{aligned} \tag{6.43}$$

where from:

$$\omega_3 = \omega_1 \frac{R_1}{R_3} = \omega_1 \frac{z_1}{z_3}. \tag{6.44}$$

We also have the relation:

$$R_1 + 2R_2 = R_3; \ (z_1 + 2z_2 = z_3). \tag{6.45}$$

Then it follows:

$$\omega_3 = \omega_1 \frac{z_1}{z_1 + 2z_2}, \tag{6.46}$$

and the transmission report will be given by:

$$i_{1,3} = \frac{\omega_1}{\omega_3} = \frac{z_1 + 2z_2}{z_1}. \tag{6.47}$$

(b) Wheel 3 fixed. In this case, the speed distribution is represented in Fig. 6.51. Based on this distribution we can write:

$$v_1 = \omega_1 R_1 = \omega_2 2R_2;$$

$$v_2 = \omega_2 R_2 = \omega_H (R_1 + R_2), \tag{6.48}$$

where from:

$$\omega_H = \frac{\omega_1 R_1}{2(R_1 + R_2)} = \frac{\omega_1 z_1}{2(z_1 + z_2)}, \tag{6.49}$$

and the transmission ratio is given by:

$$i_{1,H} = \frac{\omega_1}{\omega_H} = \frac{2(z_1 + z_2)}{z_1}. \tag{6.50}$$

(c) Wheel 1 fixed (the crank is the driving element). Based on the speed distribution (Fig. 6.51) we can write:

$$\begin{aligned} v_1 &= \omega_H (R_1 + R_2) = \omega_2 R_2; \\ v_2 &= 2\omega_2 R_2 = \omega_3 (R_1 + 2R_2), \end{aligned} \tag{6.51}$$

where from:

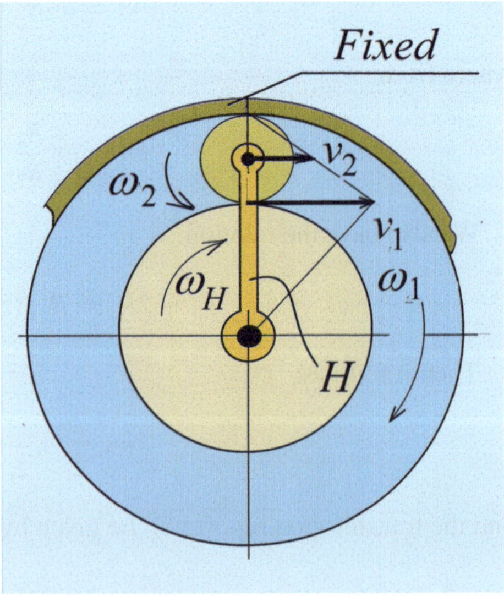

Fig. 6.51 External wheel fixed

Fig. 6.52 Central wheel fixed

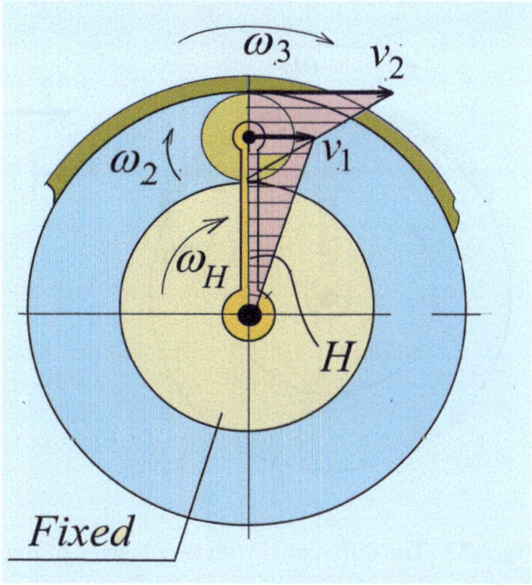

$$\omega_3 = \frac{\omega_H(R_1 + R_2)}{2(R_1 + 2R_2)} = \frac{\omega_H(z_1 + z_2)}{2(z_1 + 2z_2)}. \tag{6.52}$$

The transmission ratio is:

$$i_{H,1} = \frac{\omega_H}{\omega_1} = \frac{2(z_1 + 2z_2)}{z_1 + z_2}. \tag{6.53}$$

(d) If all the elements move then we have the actual differential mechanism.

In this case, the speed distribution is shown in Fig. 6.53. We assumed that the central wheel 1 has a rotational movement clockwise with the angular velocity ω_1 and the wheel 3 has a rotational movement with the angular velocity ω_3. We have $v_1 = \omega_1 R_1$, $v_3 = \omega_3 R_3$. Considering the figure, the similarity relations between the triangles can be written:

$$\frac{v_1}{x} = \frac{v_3}{2R_2 - x}, \tag{6.54}$$

which gives the position of the instantaneous center of rotation for wheel 2:

$$x = \frac{2v_1 R_2}{v_1 + v_3}. \tag{6.55}$$

From similarity:

$$\frac{v_2}{v_3} = \frac{R_2 - x}{2R_2 - x}, \tag{6.56}$$

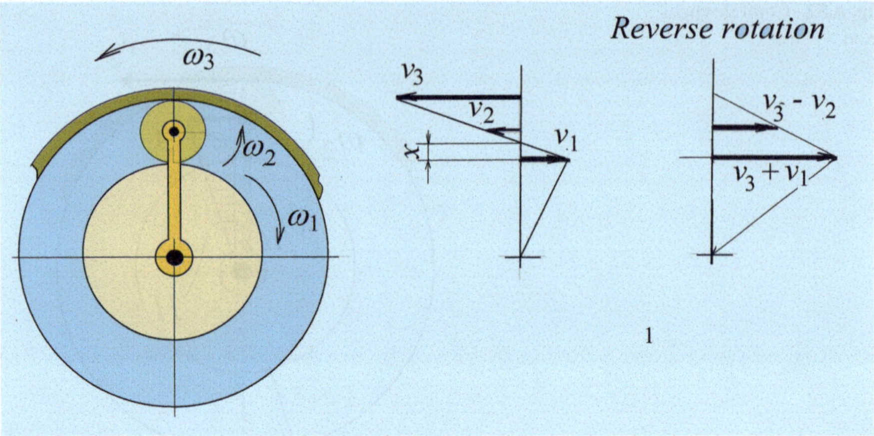

Fig. 6.53 The differential with two degrees of freedom

It results the expression of v_2:

$$v_2 = \frac{v_3 - v_1}{2}. \tag{6.57}$$

The relationship: $v_2 = (R_2 - x)\omega_2$ gives us the angular speed ω_2 of the satellite:

$$\omega_2 = \frac{v_2}{R_2 - x} = \frac{v_1 + v_3}{2R_2}. \tag{6.58}$$

I could also get ω_2 from the relationship $\omega_2 = v_1/x$. To find out the angular speed of the crank we use the relation:

$$v_2 = \omega_H(R_1 + R_2) = \frac{v_3 - v_1}{2}, \tag{6.59}$$

where from:

$$\omega_H = \frac{v_3 - v_1}{2(R_1 + R_2)}. \tag{6.60}$$

The relationship between the angular velocities of the rotating elements is given by:

$$\omega_H(2R_1 + R_2) - \omega_3 R_3 + \omega_1 R_1 = 0$$

$$\text{or } (\omega_H(2z_1 + z_2) - \omega_3 z_3 + \omega_1 z_1 = 0). \tag{6.61}$$

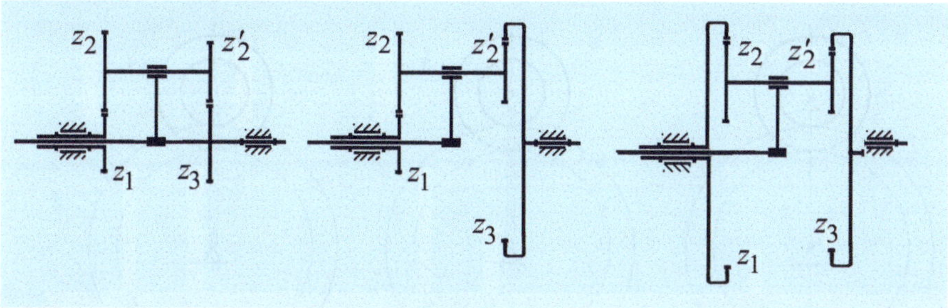

Fig. 6.54 Versions of double satellite differential

A more elegant way to obtain the solution is to give the whole mechanism a reverse motion so that one of the elements is fixed. For example, if we give a rotational movement with the angular velocity $-\omega_3$, then the gear crown will be at rest, the crank will have the angular velocity $\omega_3 - \omega_H$, the central wheel 1 will have the angular velocity $\omega_1 + \omega_3$ and the satellite two the angular velocity $\omega_3 - \omega_2$. In this case, the speed distribution becomes that of Fig. 6.50 and the solution is based on the relationships already studied.

6.3.2 *Double Satellite Differential*

In practice, the case can be encountered when we are dealing with a double satellite differential. Figure 6.54 shows some constructive cases of differentials with double satellite. For the first of these differentials, we will also do an analysis of the mode of motion transmission. The cases that can arise are:

(a) Fixed crank. The mechanism becomes a gear mechanism with fixed axes. Based on the speed distribution represented in Fig. 6.55a, it can be written:

$$v_1 = \omega_1 R_1 = \omega_2 R_2;\ v_2 = \omega_2 R_2' = \omega_3 R_3, \tag{6.62}$$

where from:

$$\omega_3 = \omega_1 \frac{R_1 R_2'}{R_2 R_3} = \omega_1 \frac{z_1 z_2'}{z_2 z_3}. \tag{6.63}$$

The transmission ratio is:

$$i_{1,3} = \frac{\omega_1}{\omega_3} = \frac{z_2 z_3}{z_1 z_2'}. \tag{6.64}$$

(b) Wheel 3 fixed. The speed distribution represented in Fig. 6.55b allows us to write:

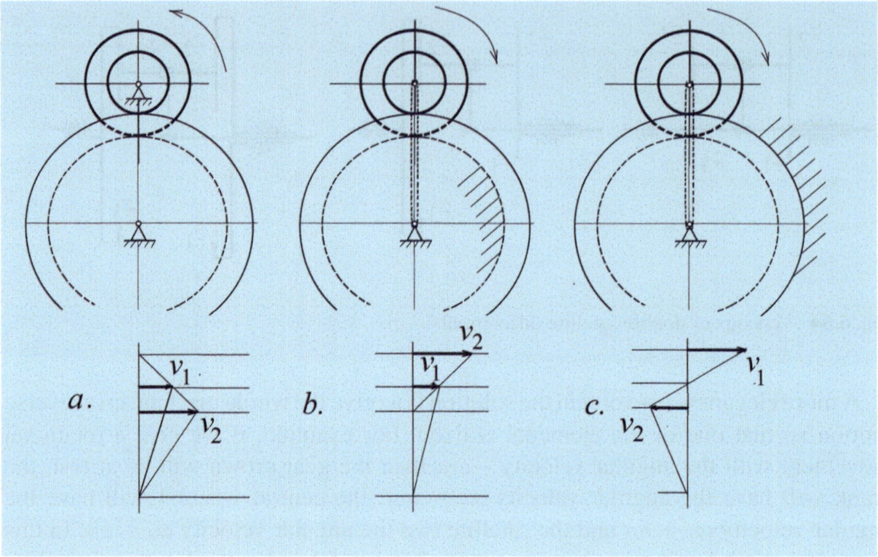

Fig. 6.55 Different versions of differential

$$v_1 = \omega_1 R_1 = \omega_2 R_2 = \omega_2 (R_2' - R_2);\ v_2 = \omega_2 R_2' = \omega_H (R_1 + R_2), \tag{6.65}$$

where from:

$$\omega_H = \omega_1 \frac{R_1 R_2'}{(R_2' - R_2)(R_1 + R_2)} = \omega_1 \frac{z_1 z_2'}{(z_2' - z_2)(z_1 + z_2)}, \tag{6.66}$$

and the transmission ratio is:

$$i_{1,H} = \frac{\omega_1}{\omega_H} = \frac{(z_2' - z_2)(z_1 + z_2)}{z_1 z_2'}; \tag{6.67}$$

$$i_{1,H} = \frac{\omega_1}{\omega_H} = \frac{(z_2' - z_2)(z_1 + z_2)}{z_1 z_2'}; \tag{6.68}$$

(c) wheel 1 fixed (Fig. 6.55c). Based on the speed distribution, we can write:

$$v_1 = \omega_H (R_1 + R_2) = \omega_2 R_2;\ v_2 = \omega_2 (R_2' - R_2) = \omega_3 R_3, \tag{6.69}$$

where from:

$$\omega_3 = \omega_H \frac{(R_1 + R_2)(R_2' - R_2)}{R_2 R_3} = \omega_H \frac{(z_1 + z_2)(z_2' - z_2)}{z_2 z_3}. \tag{6.70}$$

An application of great practical importance of the differential is in the automotive industry, equipping all the vehicles that are manufactured at the moment (Fig. 6.56).

The input torque, represented by the thicker arrow (Figs. 6.57, 6.58 and 6.59), is distributed equally to the two output shafts (thin arrows), which turn the two wheels, if the moments resisting the wheels are the same.

The horizontal satellite will not rotate.

If the left wheel meets resistance (Fig. 6.58), it will stop, the whole assembly will continue to rotate, the horizontal satellite will rotate over the left vertical satellite, fixed, and the right wheel will rotate at a double speed (results if the calculations are made) against the speed of the differential case.

The differential can also be used between two axles of a vehicle, to distribute the motion to the two axles and prevent the vehicle from skidding. Compared to the previously studied cases, the difference appears in the fact that the input and output

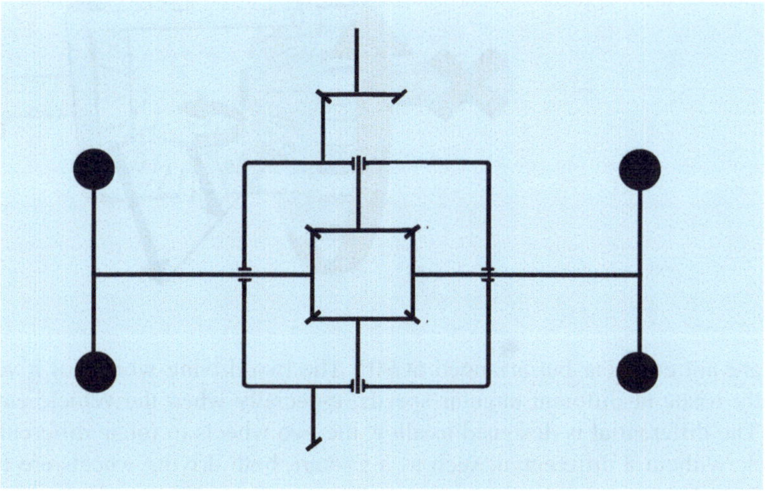

Fig. 6.56 Kinematic scheme a differential mechanism for cars

Fig. 6.57 Speed distribution at straight line motion [25]

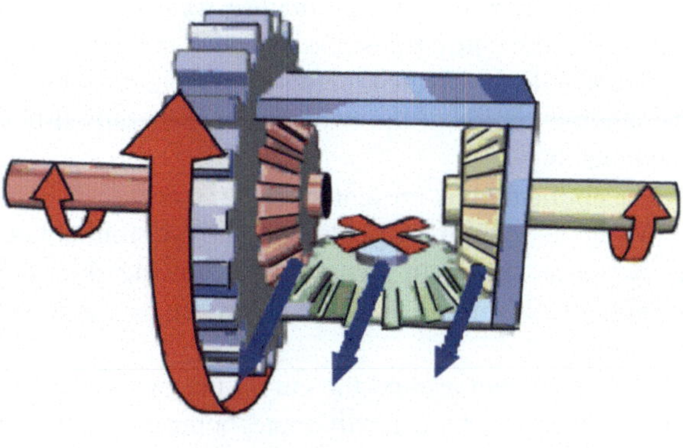

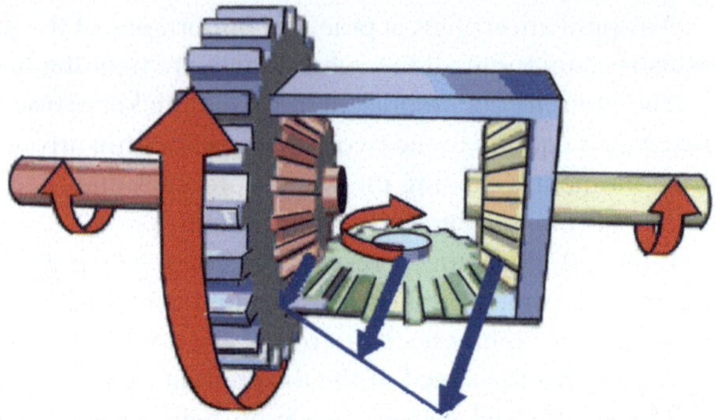

Fig. 6.58 Speed distribution at curve motion [26]

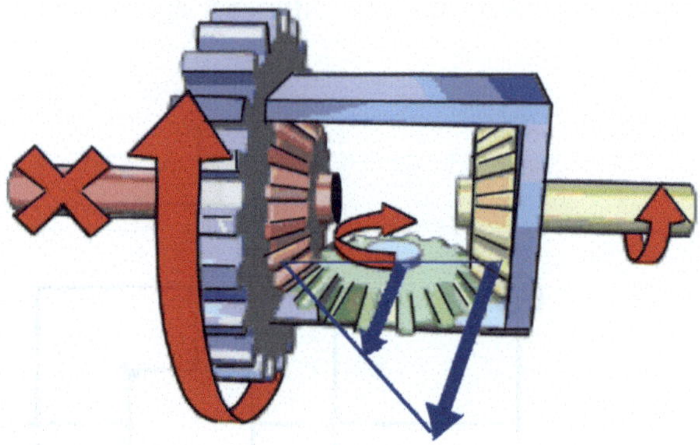

Fig. 6.59 One of the wheels stops [26]

shafts are not collinear but arranged at 90°. The two driving wheels of a vehicle generally rotate at different angular speeds, especially when the vehicle enters a curve. The differential is designed to allow the two wheels to rotate differently. In a vehicle without a differential, such as a go-kart, both driving wheels are forced to rotate at the same angular speed, generally around a common axis, by a motion transmission mechanism consisting of simple gears. When cornering, the inner wheel travels a shorter distance than the outer one, resulting in wheel slippage. This leads to unpredictable driving, premature wear of the wheels and additional stress and damage to the entire transmission mechanism.

We will deal with the kinematics part, namely the relationship that exists between the angular speeds of the two wheels (generally different) and the angular speed of the input shaft.

An application is presented in Fig. 6.60.

When moving in a straight line, the differential housing rotates, driving the planetary pinion shown in green on the drawing (Fig. 6.57). As the two wheels move identically, the pinion drawn in green behaves as if it made a common body with the housing.

When moving around the curve, the two wheels will have different speeds, and the pinion colored green will rotate, being driven by the wheels that move differently.

Fig. 6.60 An appliction of the truck differential [27]

But its center will move with the speed of the casing (Fig. 6.58). The relationship between the angular velocities of the two wheels is obtained by assuming that the speed of the center of the satellite rotating with the casing is the arithmetic mean of the speeds of its points of contact with the pinions that drive the wheels:

$$\frac{v_s + v_d}{2} = v_c, \tag{6.71}$$

where from:

$$\omega_s + \omega_d = 2\omega_c. \tag{6.72}$$

In the extreme case that one of the wheels stops, the other wheel will rotate with double angular speed (Fig. 6.59).

How the motor torque is transmitted to the two wheels is a problem that goes beyond the framework of kinematics.

It seems that the differential has been known for a very long time, in the history of humanity the following milestones can be noted (according to wikipedia) (Fig. 6.61).

- 1050 BC-771 BC: The *Book of Song* mentions the South Pole Chariot, which uses a differential mechanism, invented during the Western Zhou Dynasty.
- 150 BC-100 BC—The Antikythera Mechanism, discovered on an ancient shipwreck near the Greek island of Antikythera, uses a differential mechanism.
- 227–239 dH—Ma Jun during the Wei Dynasty in China, invents the Chinese compass (the South Pointing Chariot), which indicates the direction of the cardinal points through a non-magnetic means, using the differential mechanism.

Fig. 6.61 Chinese compass [28]

- 658, 666 dH—two Chinese Buddhist monks and engineers build the Chinese compass for Emperor Tenji of Japan.
- 1027, 1107 dH—detailed sketches of the Chinese compass (the South Pointing Chariot) by Yan Su and then Wu Deren, which describe the mechanical functions and transmission ratios in much more detail than the previous mentions.
- 1720—Joseph Williamson uses a differential clock mechanism.
- 1810—Rudolph Ackermann of Germany invents a four-wheel steering system for carriages, which was later erroneously recorded as a differential mechanism
- 1827—the modern automobile differential was patented by clockmaker Onésiphore Pecqueur (1792–1852) of the Conservatoire des Arts et Métiers in France for use on a steam-powered carriage.
- 1832—Richard Roberts from England patents 'gear of compensation', a differential for locomotives.
- 1876—James Starley of Coventry invents a differential for bicycle use; invention later used in automobiles by Karl Benz.
- 1897—first use of differential on a steam carriage in Australia by David Shearer.
- 1913—Packard introduces the helical gear differential, which reduces noise.
- 1926—Packard introduces the differential with hypoid teeth, with major constructive advantages.

References

1. Schiehlen W, Eberhard P (2006) Technische Dynamik. Teubner, Modele für Regelung und Simulation
2. Teodorescu PP (2006) Mechanical systems, classical models. Springer Verlag
3. Vlase S, Teodorescu PP, Itu C, Scutaru ML (2013) Elasto-dynamics of a solid with a general "rigid" motion using fem model. Part II: Analysis of a double cardan joint. Romanian J Phys 58(7–8):882–892
4. Vlase S, Marin M, Öchsner A (2019) Considerations of the transverse vibration of a mechanical system with two identical bars. Proc Inst Mech Eng Part L-J Mater-Des Appl 233(7):1318–1323. https://doi.org/10.1177/1464420717745109
5. Vlase S, Negrean I, Marin M, Năstac S (2020) Kane's method-based simulation and modeling robots with elastic elements, using finite element method. Mathematics 8(5):805. https://doi.org/10.3390/math8050805
6. Stanciu MD, Vlase S, Marin M (2019) Vibration analysis of a guitar considered as symmetrical mechanical system. Symmetry 11(6):727. https://doi.org/10.3390/sym11060727
7. Scutaru ML, Vlase S, Marin M (2023) Symmetrical mechanical system properties-based forced vibration analysis. J Comput Appl Mech 54 (4):501–514 https://doi.org/10.22059/JCAMECH.2023.366509.890.
8. Scutaru ML, Vlase S (2012) Some properties of motion equations describing the nonlinear dynamical response of a multibody system with flexible elements. J Appl Math 1:628503. https://doi.org/10.1155/2012/628503
9. Vlase S, Marin M, Öchsner A (2019) Considerations of the transverse vibration of a mechanical system with two identical bars. Proc Inst Mech Eng Part L: J Mater: Des Appl 233(7):1318–1323. https://doi.org/10.1177/1464420717745109.
10. Stanciu E, Bencze A, Cazan A, Vlase S, Marin M, Öchsner A (2025) Effects of infill parameter variations on mechanical properties in 3D printed structures: comparative study with FEM analysis. J Comput Appl Mech 56(2):318–330. https://doi.org/10.22059/jcamech.2025.390563.1371
11. Cornell University Library, Digital Library Collections. Available at https://digital.library.cornell.edu/catalog/ss:372768
12. Cornell University Library, Digital Library Collections. Available at https://digital.library.cornell.edu/catalog/ss:372774
13. Masana A. Mechanical systems illustrations. Available athttps://albertmasana.wordpress.com/2012/04/04
14. Science Photo Library, Raising a slab of stone using a block and tackle mechanism. Available at https://www.sciencephoto.com/media/1002623/view/raising-a-slab-of-stone-using-a-block-and-tackle-mechanism
15. MT24, Worm gear—mechanical assemblies. Available at https://www.mt24.com/en/mechanics/mechanical-assemblies/worm-gear.html?b=googlemerchant&gad_source=1&gad_campaignid=20984519506&gbraid=0AAAAAqvB_-bqf9Z0NosnDmJ4elaUsZqwm&gclid=Cj0KCQiApL7KBhC7ARIsAD2Xq3A8vdzV32fWCdAskgJY_MmZlBTytRUxslvTxslv_FlwCSGsb6HNaZsaAhsUEALw_wcB
16. https://www.emag.ro/cremaliera-metalica-zincata-bft-cvz-s-pentru-automatizari-porti-culisante-1m-bft-cvz-s/pd/DNT9SHMBM/
17. Passei Direto, Types of gears used. Available at https://www.passeidireto.com/arquivo/37045347/tipos-de-engrenagens-utilizadas-1
18. Ritacco Automotive, Automotive transmission systems. Available at https://www.ritacco-automotive.eu/
19. TradeIndia, Heavy industrial gear box. Available at https://www.tradeindia.com/products/heavy-industrial-gear-box-118103.htmL
20. China Reducers, Spiral bevel gears. Available at https://www.china-reducers.com/spiralbevelgears.htm

21. Europeana Collections, Historical mechanical image. Available at https://www.europeana.eu/en/item/2020801/dmglib_handler_image_4782023
22. Dudiţă F (1964) Transmisii cardanice. Ed. Tehnică, Bucureşti
23. Morin D (2008) Introduction to classical mechanics: with problems and solutions, 1st edn. Cambridge University Press
24. Bancroft Library, University of California, Berkeley, Early physics instruments. Available at http://bancroft.berkeley.edu/Exibits/physics/origins02.html
25. https://ro.wikipedia.org/wiki/Fi%C8%99ier:Differential_free.png
26. Wikimedia Commons, Locked differential mechanism. Available at https://commons.wikimedia.org/wiki/File:Differential_locked.png
27. AstroSafari Forum, Differential cross-pin retaining bolt discussion. Available at https://www.astrosafari.com/threads/rounded-head-on-differential-cross-pin-retaining-bolt.152898
28. Softpedia, How did the compass appear? Available at https://news.softpedia.com/news/How-Did-the-Compass-Appear-77628.shtml26

The manufacturer's authorised representative in the EU is Springer Nature Customer Service Centre GmbH, Europaplatz 3, 69115 Heidelberg, Germany. If you have any concerns regarding our products, please contact ProductSafety@springernature.com

Printed and bound by CPI Group (UK) Ltd, Croydon, CR0 4YY

05/08/2026

02186827-0001